AF745588

Studies in Autonomic, Data-driven and Industrial Computing

Series Editors

Swagatam Das, Indian Statistical Institute, Kolkata, India

Jagdish Chand Bansal, South Asian University, Chanakyapuri, India

The book series Studies in Autonomic, Data-driven and Industrial Computing (SADIC) aims at bringing together valuable and novel scientific contributions that address new theories and their real world applications related to autonomic, data-driven, and industrial computing. The area of research covered in the series includes theory and applications of parallel computing, cyber trust and security, grid computing, optical computing, distributed sensor networks, bioinformatics, fuzzy computing and uncertainty quantification, neurocomputing and deep learning, smart grids, data-driven power engineering, smart home informatics, machine learning, mobile computing, internet of things, privacy preserving computation, big data analytics, cloud computing, blockchain and edge computing, data-driven green computing, symbolic computing, swarm intelligence and evolutionary computing, intelligent systems for industry 4.0, as well as other pertinent methods for autonomic, data-driven, and industrial computing.

The series will publish monographs, edited volumes, textbooks and proceedings of important conferences, symposia and meetings in the field of autonomic, data-driven and industrial computing.

Haribhau R. Bhapkar · Parikshit N. Mahalle

Discrete Mathematics: An Experiential Learning Perspective

Springer

Haribhau R. Bhapkar
Department of Mathematics
Central University of Kashmir
Tulmulla, Jammu and Kashmir, India

Parikshit N. Mahalle
Department of Artificial Intelligence
and Data Science
Vishwakarma Institute of Technology
Pune, Maharashtra, India

ISSN 2730-6437 ISSN 2730-6445 (electronic)
Studies in Autonomic, Data-driven and Industrial Computing
ISBN 978-981-95-7209-0 ISBN 978-981-95-7210-6 (eBook)
https://doi.org/10.1007/978-981-95-7210-6

© The Editor(s) (if applicable) and The Author(s), under exclusive license to Springer Nature Singapore Pte Ltd. 2026

This work is subject to copyright. All rights are solely and exclusively licensed by the Publisher, whether the whole or part of the material is concerned, specifically the rights of translation, reprinting, reuse of illustrations, recitation, broadcasting, reproduction on microfilms or in any other physical way, and transmission or information storage and retrieval, electronic adaptation, computer software, or by similar or dissimilar methodology now known or hereafter developed.
The use of general descriptive names, registered names, trademarks, service marks, etc. in this publication does not imply, even in the absence of a specific statement, that such names are exempt from the relevant protective laws and regulations and therefore free for general use.
The publisher, the authors and the editors are safe to assume that the advice and information in this book are believed to be true and accurate at the date of publication. Neither the publisher nor the authors or the editors give a warranty, expressed or implied, with respect to the material contained herein or for any errors or omissions that may have been made. The publisher remains neutral with regard to jurisdictional claims in published maps and institutional affiliations.

This Springer imprint is published by the registered company Springer Nature Singapore Pte Ltd.
The registered company address is: 152 Beach Road, #21-01/04 Gateway East, Singapore 189721, Singapore

If disposing of this product, please recycle the paper.

Preface

Yogaḥ karmasu kauśalam—Excellence in learning arises through action.

—Bhagavad Gita 2.50

Mathematics is often perceived as abstract, symbolic, and distant from tangible experience. Yet, when viewed through the lens of doing, discovering, and reflecting, it becomes a powerful tool for nurturing reasoning, creativity, and problem-solving. This book, *Discrete Mathematics: An Experiential Learning Perspective*, has been written with the conviction that learners understand mathematical concepts best when they engage with them through action, experimentation, and application.

Mathematics is the language of science and technology and research. Discrete mathematics forms the backbone of computer science, artificial intelligence, data analytics, and network design. It provides the logic, structure, and precision required to design algorithms, optimize systems, and interpret information. However, the teaching of discrete mathematics traditionally leans heavily on theory. This book aims to bridge that gap by introducing an experiential dimension—learning by doing, reflecting, and connecting theory with practice. There are two types of project-based learnings. One is to apply knowledge to deepen the understanding, and the other is to understand or define the concept or theory. This book follows the **"BEIMI"** philosophy or theory to define the concepts. The **"BEIMI"** stands for Brainstorming Experiments/Activities (BE), Inferences (I) and Mathematical Insights (MI). Moreover, the case studies/ projects are given chapter wise.

Each chapter of the book integrates hands-on activities, visual explorations, real-world applications, and case studies to promote deeper engagement. The design emphasizes constructivist pedagogy, where learners build understanding through inquiry and experimentation. By embedding experiential learning cycles—Concrete Experience, Reflective Observation, Abstract Conceptualization, and Active Experimentation—this text encourages students to move beyond rote problem-solving toward conceptual mastery and creative application.

Chapter 1 sets the stage by explaining why discrete mathematics is essential for engineers and scientists, presenting its novel philosophy and broad applications. Chapter 2 introduces the idea of experiential learning itself—how learning occurs through doing, exploring, and reflecting—and how this pedagogy can transform the study of discrete mathematics.

Chapters 3–13 cover core topics—Sets, Relations, Functions, Logic, Counting, Probability, Graphs, Trees, Generating Functions, and Algebraic Structures—each approached through interactive learning, visualization, and application-based exercises. Every concept is linked with case studies, projects, and computational implementations, enabling learners to experience mathematics as an evolving, applicable discipline rather than a static body of facts.

The concluding Chap. 14: Conclusion and Future Outlook synthesize the learning journey, discusses open research directions, and reflects on how experiential education in discrete mathematics can shape innovative thinkers and practitioners in science and engineering.

The pedagogical approach of this book aligns with the educational vision of *learning by doing*, rooted in both ancient wisdom and modern learning science. As the *Bhagavad Gita* teaches, **"Yogaḥ karmasu kauśalam"—Excellence is achieved through action**. True mathematical understanding, likewise, is cultivated through purposeful engagement with problems, not through passive observation.

It is hoped that this book will serve as a guide and inspiration for teachers, learners, and researchers who wish to experience mathematics not merely as a subject, but as a living process of exploration and innovation. By blending conceptual clarity with experiential learning, this work aspires to nurture a generation of learners who can think critically, act creatively, and apply mathematics meaningfully in a rapidly evolving technological world.

Tulmulla, India Dr. Haribhau R. Bhapkar
Pune, India Dr. Parikshit N. Mahalle

Competing Interests The authors have no competing interests to declare that are relevant to the content of this manuscript.

Contents

Chapter 1
Why Discrete Mathematics

Like the crest of a peacock, like the gem on the head of a snake,
so is Mathematics at the head of all knowledge.

Abstract Mathematics is often perceived as abstract, symbolic, and distant from tangible experience. Hitherto, when viewed through the lens of doing, discovering, and reflecting, it has become a powerful tool for nurturing reasoning, creativity, and problem-solving. This chapter deals with the novel philosophy and application of Discrete Mathematics. The authors focus with the conviction that learners understand mathematical concepts more precise when they engage with them through action, experimentation, and application. The importance of Discrete Mathematics and Mathematics is defined as "Like the crest of a peacock, like the gem on the head of a snake, so is Mathematics at the head of all knowledge." The chapter concludes with various applications of Discrete Mathematics in Science, Technology, Mathematics and real life.

Keywords Discrete mathematics · Experiential learning · Service learning · Scientific approach

1.1 Introduction

Performing experiments to gain some experience is a preferred practice in the social, personal, professional, and technical world. This process of learning by doing or through some experiences is known as Experiential Learning (EL) which was initially introduced by John Deway and Jean Piaget [1, 2]. However, the credit of EL is with David A. Kolb [3] who has made it more popular and made it accepted to the technical community. Kolb has introduced the foundational theory of EL proposing the process of learning through experiences toward knowledge transformation. As per Kolb, learning happens due to the creation of knowledge and knowledge is created based on experiences. The better we transform our experiences, the better the knowledge

© The Author(s), under exclusive license to Springer Nature Singapore Pte Ltd. 2026
H. R. Bhapkar and P. N. Mahalle, *Discrete Mathematics: An Experiential Learning Perspective*, Studies in Autonomic, Data-driven and Industrial Computing,
https://doi.org/10.1007/978-981-95-7210-6_1

created is. In the practical world, there are two types of experiences which include direct experience and indirect experience and are listed below

- **Direct Experience**: Directly gained by an individual, i.e., personal experience. For example, a kids experience of drinking hot milk and he feels mouth burn feeling which is experienced with a kid directly.
- **Indirect Experience**: Gained from recommendation and experiences of others. For example, somebody is telling a kid to drink milk carefully as it is hot so that the kid can avoid the experience of mouth burning. Indirect experience refers to the knowledge or action initiated by someone's recommendation or experience.

Both direct and indirect experiences play an important role in enhancing the learning curve. Particularly for subjects like Discrete Mathematics (DM), connecting theoretical concepts with the real-world practical use cases makes this mathematical subject more convincing. Deeper and comprehensive understanding of DM makes learners more competent in applied knowledge. EL focuses more on combining direct experience with the underlined reflection using past knowledge and experience. Reflection helps learners to understand whether the outcome of learning is useful or not. EL also enables active involvement of both learner and mentor and encourages more collaboration and exchange of ideas as well as perspectives. The key benefits of EL are listed below:

Critical Thinking: EL enhances critical thinking by analyzing the positive and negative sides of the situation. It also helps to decide whether the outcomes are beneficial or not.
Problem Solving: EL also makes problem-solving process easier and simpler as the underlined situation is deeply analyzed.
Contextual Decision-Making: Situational decision-making skills is an important outcome of EL.
Ideas and Skills: EL helps to explore more ideas and skills as every problem has been understood with deeper insights into critical thinking. This also led to more opportunities.
Reflection: EL helps to develop the ability of reflection making learner understand the usability of outcomes.

There are various types of EL and are presented in Fig. 1.1.

Comparative education helps to compare different types of education received from experiences. Comparative education also helps to differentiate between different types of subjects and domains. In internships, all theoretical learning is practically applicable in the field for certain duration and more practical insights are gained. All teaching methods and strategies are taken into consideration by learners in internships. Practicum is one type of EL where the problem is divided into smaller tasks and these tasks are designed, developed and implemented in the laboratory. Service learning is specifically applicable to the professionals where they learn by explorations. Previous experiences are also incorporated into the learning in service-learning type of EL. In research type of EL, we explore better solutions as compared to the previous by adding some innovation or novelty. Cooperative learning helps us

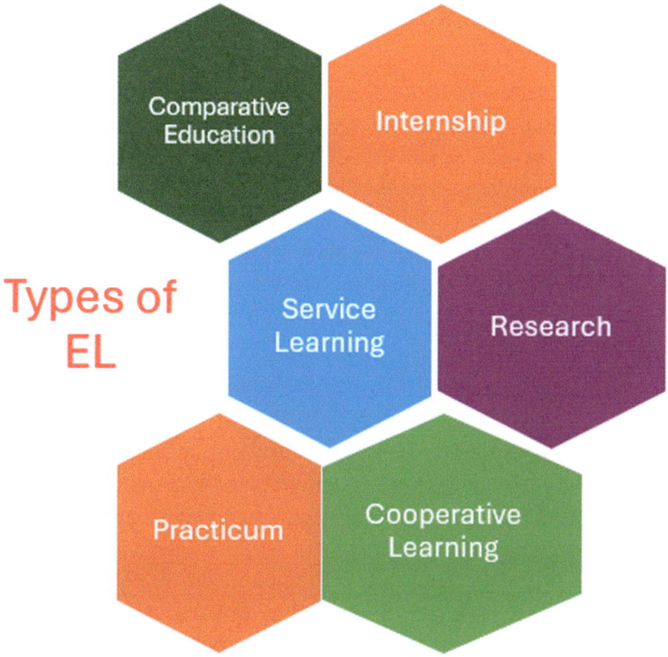

Fig. 1.1 Types of EL

to work in a team where we collectively decide which tools and methods to apply for learning.

In EL, the learning process depends on the various factors and does not always start with the experience. It is the learner's responsibility to decide which mode of learning to select based on the underlined domain and situation. Let's consider the real-world example of learning driving.

- Actor: Person
- Objective: Learning to drive car
- Possible modes of learning
  - Reflection: Learning car driving by observing other people driving.
  - Reading Instruction Book: Some learners might select to learn car driving by reading the instruction book.
  - Driving on Test Course: Some learners might prefer to learn by seating behind the seat of the car and participating in driving on a test score.

The above example shows that EL develops the ability to immediately apply the knowledge to solve the problem, enables access to real-time feedback, helps to promote teamwork and improve communication skills. EL also helps to develop high order thinking skills and reflective practice habits.

1.2 Novel Philosophy of Discrete Mathematics

With the increasing demands of users on the internet, today we are in the era of high-performance computing applications. User experience, particularly in the form of response time, is increasingly demanding. Computer science, information technology, and artificial intelligence and allied topics are playing a crucial role in this era of industry 4.0 and society 5.0 [4, 5].

The novel philosophy of DM is to apply the fundamental concepts of solving problems in broader areas of engineering and sciences and is depicted in Fig. 1.2. Mainly two broad areas which include scientific problems and societal problems and are discussed below in detail.

1. **Scientific Problems**

DM being a specialized branch of computer science is specifically used in the algorithms, programming languages and cryptography. Automated theorem proving, building of software development, design and development of cryptographic algorithms. Important concepts and mathematical notations derived from the DM play vital role in the field of algorithmics and some intensive applications in real world. Emerging topics like network modeling, design and analysis, computer optimization problems, network optimization problems, and analysis of data structures in the algorithmic perspective are heavily dependent on the graph theory and combinatorial structures. Most of the real problems are based on graph theory and used to address complex and practical problems. In addition to applications of DM in engineering and sciences, interdisciplinary areas like biochemistry, VLSI design, Agri-tech, fintech also using various tools and techniques of DM and presenting diverse scientific areas where DM can be used for design and modeling. Use of DM and allied concepts in modeling biological systems like genetic computational networks, ecological interactions are emerging applications in the current paradigm. Use cases like building simulation platform for species interaction, genes interaction for translational medical research uses core concepts of DM toward medical research.

Mathematical modeling of the problem is very important in research and development. Representing a given computational problem using tools and language of mathematics for the unambiguous interpretation of the problem is well received in the research. Fundamental concepts of mathematics like set theory, relations and

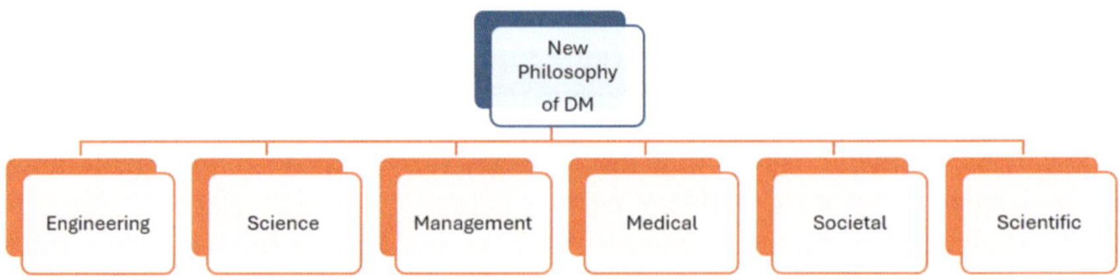

Fig. 1.2 New philosophy of DM

functions are useful in modeling the given problem [6]. With the emergence of artificial intelligence (AI), data-driven algorithms are transforming at a faster rate in terms of scale and speed. All these AI driven algorithms are based on the core concept of DM for analyzing large datasets. Probability and statistics from DM are the basis for understanding data tendency, how the data is distributed and its skewness. All machine-learning classification and clustering algorithms are based on statistics and are used to perform predictions using the discrete data points. Consider the example of unsupervised machine-learning techniques which work on unlabeled data points and the key operations include association and clustering. Clustering is fundamentally a data-mining technique which groups unlabeled data into one group based on some similarities or differences. Association mining or dimensionality reduction is some of the examples in unsupervised machine learning. These machine-learning techniques are more appropriate for raw and unknown data as the labels are not associated with the data points. Hierarchical clustering, K-means and neural networks are the most popular algorithms in this category. In cluster analysis or clustering, unlabeled data is taken as input, and it groups the set of data points in such a way that data points or objects in the same group, i.e., cluster are more similar based on some properties or parameter than the objects in the other group. Pattern recognition, image analysis and exploration, and information retrieval are some important applications of unsupervised machine learning. Consider the emerging use case of autonomous vehicles and it is described below:

Use case:
Autonomous vehicle/Driverless car

Type of Machine Learning:
Unsupervised machine learning

Dataset:
Unlabeled

Operation:
Clustering

Description:
In this case, unsupervised machine-learning algorithms on the fly learn new road features, and there is very minimal input from a human or driver during the process. Unsupervised machine-learning algorithm address the problem of addressing unlabeled data and learning this data toward classification. Later, the labels are added to the dataset which becomes easier and more convenient. Like this, this technique is also useful in identifying patterns in the data which cannot be addressed using other normal methods.

DM focuses more on structures which are discrete in nature than continuous structures. Ontological and epistemological implications are two main factors of DM and are described below:

- **Ontological Perspective**: Deals with the criteria for any structure to be discrete and how discreteness is connected to the real-world problems. There are many

emerging areas like quantum machines and high-performance computing systems which are majorly based on discrete structures.
- **Epistemological Perspective**: Deals with the evolution of discrete structures and the tradeoff between discrete and non-discrete systems and its impact of the real-world problem solving and solution.

DM and logic development goes hand in hand. Main components of DM which include truth tables, predicate and propositional calculus are based on the foundations of logic. DM with experiential learning needs to explore the following three important parameters

- **Computation**: Deals with the role of DM in the design of efficient computationally intensive processes and algorithms.
- **Complexity**: Deals with the role and impact of DM in analyzing the complexity of natural and artificial use cases and systems.
- **Logic**: Deals with the use of appropriate logic like symbolic logic or discrete logic and their use to address real-world problems.

In view of this, broad application areas of DM are discussed in the next section.

1.3 Applications of Discrete Mathematics

DM is used in many broad areas of computer science and spans to everyday applications. Applications which consist of mainly discrete systems and discrete structures as key functional components are potential applications of DM and are depicted in Fig. 1.3 as well as discussed below.

1. **Computer Science and Algorithms**

Core computer science, design, and analysis of algorithms are the core application areas DM. DM is used in graph theory applications like network design, modeling, and analysis, Social media design and analysis, all shortest path algorithms in graphs are some main applications of DM. Data structures and algorithms and main data structures like trees, graphs, hash tables, and heaps are based on the core concepts of DM. Design and analysis of compilers, design and analysis of finite state machine-like nondeterministic finite automata, deterministic finite automata are also based on the DM concepts.

2. **Combinatorial Optimization**

Combinatorial optimization techniques are used to find the optimal solution from small and finite problem space. Scheduling problems in the real production floor where the task is to optimize some operations is an important application of DM. Scheduling problems using DM are popular in manufacturing and transportation industries. Optimization of network resources, routing optimization, efficiency improvements in communication protocol are based on the core functionalities of

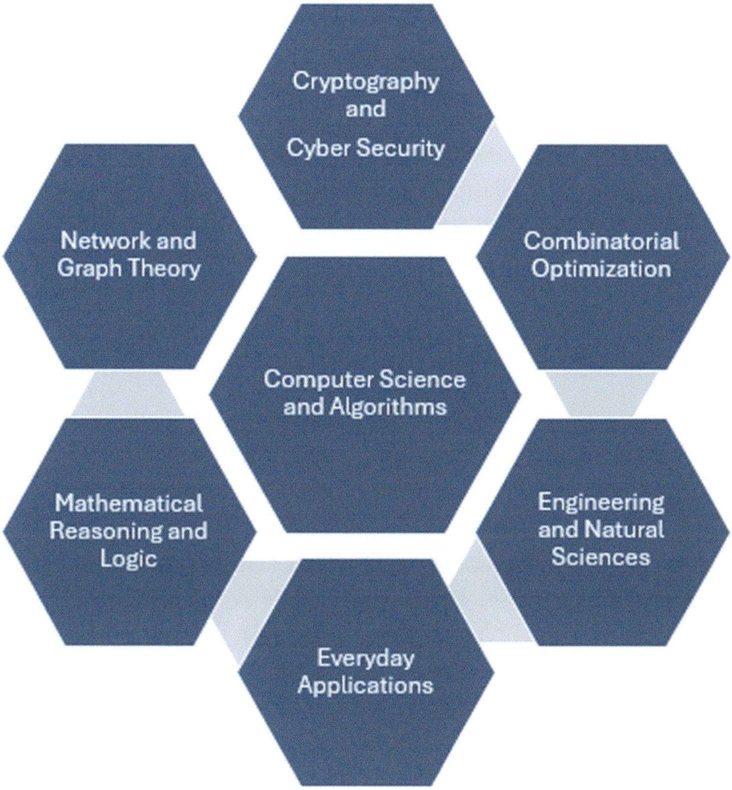

Fig. 1.3 Applications of discrete mathematics

DM. Resource optimization and finding optimal solutions which either minimize something or maximize something is based on DM.

3. **Cryptography and Cybersecurity**

Core concepts of DM like groups and rings are mainly used in designing cryptographic algorithms and techniques in the crypto world. Key management techniques like public key cryptography are based on the number theory and discrete logarithmic problems as well as modular arithmetic. Hashing techniques are based on the check sum and used to detect data tampering and used in blockchain platforms. All popular cryptographic algorithms are based on the concepts of groups and rings in DM and popular algorithms like RSA and ECC are designed using discrete structures.

4. **Network and Graph Theory**

Life sciences and analysis of biological network structures use DM concepts for analysis and understanding protein synthesis, sequencing, etc. Circuit making, design, modeling and analysis of electrical circuits are also based on the concepts of DM.

Social network analysis, relationship network analysis, influence analysis techniques in social media are designed using DM concepts.

5. **Mathematical Reasoning and Logic**

Soft computing principles are the foundations for artificial intelligence and these principles are used in forecasting, estimation, and predictions. Techniques like fuzzy logic, artificial neural networks, and evolutionary computing are based on the concepts of DM and used in logical reasoning and decision making. Formal methods based on DM are used in software engineering, especially for software verification. Formal reasoning for the proofs is also routed on the concepts of DM.

6. **Engineering and Natural Sciences**

Modeling and simulation of discrete systems, navigation of aerospace objects, their control, DNA synthesis are some emerging applications of DM.

7. **Everyday Applications**

In addition to scientific applications, concepts of DM are also used in accounting, finance, gaming industry and E-commerce. Mainly DM concepts are used for recommendation, risk analysis, and design of learning systems.

References

1. Miettinen R (2000) The concept of experiential learning and John Dewey's theory of reflective thought and action. Int J Lifelong Educ 19(1):54–72
2. Ord J, Leather M (2011) The substance beneath the labels of experiential learning: the importance of John Dewey for outdoor educators. J Outdoor Environ Educ 15(2):13–23
3. Bergsteiner H, Avery GC, Neumann R (2010) Kolb's experiential learning model: critique from a modelling perspective. Stud Contin Educ 32(1):29–46
4. Patil RV, Ambritta PN, Mahalle PN, Dey N (2022) Medical cyber—physical systems in society 5.0: are we ready? IEEE Trans Technol Soc 3(3):189–198. https://doi.org/10.1109/TTS.2022.3185396
5. Neustein A, Mahalle PN, Joshi P, Shinde GR (eds) (2023) AI, IoT, big data and cloud computing for industry 4.0, 1st edn. Springer International Publishing, Cham, Switzerland
6. Mahalle PN, Nancy Ambritta P, Sakhare SR, Kulkarni AP (2023) Foundations of mathematical modelling for engineering problem solving. Springer, Singapore

Chapter 2
Introduction

Mathematics is the language with which God has written the universe. Moreover, a single formula can open a thousand doors in mathematics—Srinivasa Ramanujan

Abstract This chapter deals with the concepts of the Experiential Learning, Learning Pedagogies, and Discrete Mathematics in Engineering applications. It also explains four pedagogical approaches which include Behaviorism, Constructivism, Social Constructivism, and Liberationist. The authors define experiential learning through 10 C models as Concept, Cognizance, Case, Connect, Critics, Comment, Cross, Converge, Conclude and Check. Moreover, the chapter describes the three stages of the "BEIMI" philosophy, that is Brainstorming Experiment (BE), Inferences (I), Mathematical Insights (MI). The chapter concludes with the applications of Discrete Mathematics and experiential learning for meaningful understanding of mathematical theory.

Keywords Experiential learning · Learning pedagogy · 10 C model · BEIMI philosophy · Engineering

2.1 Learning Curve

Teaching, learning and outcome-based education is top priority [1, 2]. National accreditation bodies like National Board of Accreditation [3] and other similar national and international bodies aim to provide maximum weightage to the outcome and learning curve is very important to measure the outcomes. The education industry is considered as a service industry and main stakeholders of this industry are students, parents, and employers. A learning curve is an attempt to visualize how someone's proficiency is related to the number of skills or experience they carry. Naturally, it is evident that proficiency is always on a growing scale with increased skillsets or experience. This concludes that the learner becomes more productive if the learning curve is better and there should be more effort on improving the learning curve.

© The Author(s), under exclusive license to Springer Nature Singapore Pte Ltd. 2026

H. R. Bhapkar and P. N. Mahalle, *Discrete Mathematics: An Experiential Learning Perspective*, Studies in Autonomic, Data-driven and Industrial Computing,
https://doi.org/10.1007/978-981-95-7210-6_2

Cumulative time in the learning curve is important performance metric and mathematically, this cumulative time T is represented as below:

$$T = t * N^m, \tag{2.1}$$

where

T Total amount of time required
t Time to perform first operation or task
N Total number of operations or tasks performed
m Slope of underlined learning curve.

A learning curve is always expressed in percentage and there is always scope for improvement. Although if the percentage of learning curve is 100%, there is still the scope of improvement in terms of quality which is completely subjective in nature. There are many applications where the learning curve cannot be measured. For example, learning a topic which is more challenging, building riding skills, singing skills which does not have visual representation or mathematical expression to represent the progression in learning curve. The applications where the mathematical representation is possible to identify the learning progression is referred to as measured learning curve. Equation (2.1) shows that learning curve is incremental in nature and the repetition of task or operations always results in better performance. There are different methods of learning in literature and are explored in the next section of this chapter. However, it is important to take note that learning curve is a very important tool to design and develop more effective learning strategies and techniques so that learning percentage is more and qualitative in nature. In the sequel, there should also be a measure of quality. S-curves are mainly used task tracking and efforts estimation and the relative progress of the task can be monitored.

S-curve is an important mathematical modeling tool used in the evaluation of project tracking and efforts estimation in terms of deadline and progress tracking. Cost monitoring and resource utilization is also important usage of S-curve. This performance tracking using S-curve is directly related to the learning curve of an individual during the process. S-curve is a modeling technique to enable data-driven decision making to optimize the learning curve. The S-curve consists of mainly three components which include baseline, target, and actual. The baseline component shows the expected outcome, the target S-curve evaluates the actual performance and actual S-curve provides the actual progress as well as comparison with baseline and target.

However, it is important to explore strategies to optimize the learning curve to enhance proficiency. Figure 2.1 depicts these strategic points for learning curve optimization.

As presented in Fig. 2.1, the selection of appropriate problem domain and related problem statement, analyzing underlined problem so that one can decide on the suitable design and algorithmic strategy to solve the problem, incremental and repeated learning methodology to incorporate change management, time window with respect to the deadline, monitoring of the continuous improvement and loopback mechanism

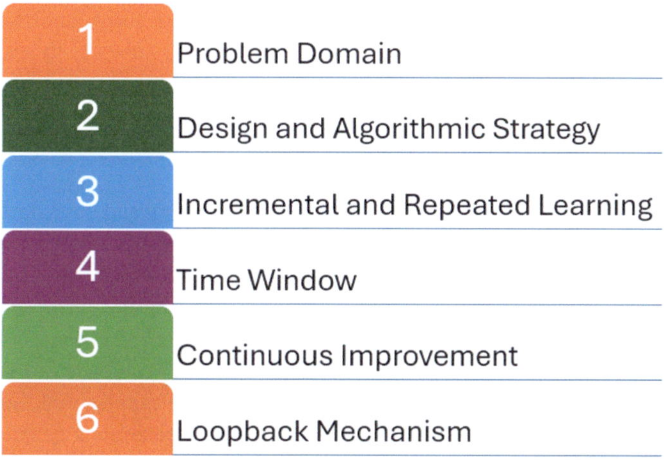

Fig. 2.1 Strategic points for learning curve optimization

to incorporate the feedback during the process are important strategic points for optimizing the learning process. There are two approaches to understand and evaluate the performance of learning curve which includes qualitative and quantitative approaches. Mathematical modeling is one of the important and useful approaches to understand learning curve performance. Mathematical components and tools play an important role in quantifying the outcome of learning curve in percentage. The language of mathematics makes this process more precise and concise resulting in the steeper slope of learning curve. Algebra and geometry, calculus, trigonometry, and number system are key mathematical components to understand learning curve more precisely.

2.2 Learning Pedagogies

Pedagogy is a very useful and popular terminology in the process of teaching and learning process which talks about methods and techniques to ensure quality teaching, qualitative and quantitative learning. Every sector in education adopts a different culture for teaching and learning process. It is very crucial to understand the relationship between adopted culture and learning methods. Learning styles and learning curve matching is an important paradigm to build skills and attitude. Pedagogy adopts a teacher-centered approach where the teacher is considered as a key entity at the center in the process of learning. The learner is always dependent on the teacher and in the sequel the teaching pedagogy used by the teacher is directly related to the learning curve. The natural assumption in the process is that the learner is willing to learn and is in the receptive mode with the clear motivation of learning. In the knowledge transfer ecosystem, the set of subjects are generally divides into three categories which includes fundamentals, conceptual and applied. Pedagogy enables a teacher-centered approach which is more suited for fundamental subjects to impart basic knowledge and concepts by using structured process. Subjects such languages and discrete mathematics (DM), which is the focus of this book requires more focus on the fundamental knowledge so that learner can apply these concepts to the real-world business cases.

In recent years, there has been lot of attention in designing and developing various national education frameworks to explore and discuss various teaching and learning pedagogies essentially for higher education. National Education Policy (NEP) 2020 [4, 5] is being well received at national level for meeting the needs of outcome and need-based education essentially in the Indian context. The main aim of NEP 2020 is to extend existing value-based education to match with today's technology-driven process. This will help to make education more experiential, and learner centered. Learner-centric approach and teacher at central position is the main philosophy of NEP 2020. Autonomy in pedagogy selection, 360 degree of holistic evaluation approach, freedom in content designing and delivery are the key components of NEP 2020.However, the success of NEP implementation highly depends on the understanding of various pedagogies.

Various pedagogical practices are closely related to the learning curve and therefore selection of appropriate pedagogy plays important role in the success of NEP. The key objectives of pedagogical approaches are presented in Fig. 2.2. There is key four pedagogical approaches which include Behaviorism, Constructivism, Social Constructivism, and Liberationist and are shown in Fig. 2.3 and discussed below.

1. **Behaviorism**

This is one of the most popular teacher-centric leaning pedagogy where the teacher is the main controller of the class and is responsible for topic-based learning. A teacher is free to use any of the modes of lesson which include explanation, teaching, modeling, and simulation, etc. A behaviorist pedagogical approach is a traditional method of teaching where the learner and teacher can be at center as per the requirement.

2. **Constructivism**

In constructivism pedagogy, the learner is at the center, and it enables experiential learning methodology. Learning based on the project activity and inquiry-based learning are some key methods adopted in constructivism pedagogy. Progressive wat of teaching through experience is the key principle used in constructivism pedagogy. This pedagogy is also referred to as invisible pedagogy being the learner at the center. This pedagogy is further explored in the scope of this book.

3. **Social Constructivism**

Social constructivism is based on the cognitive theory of learning which uses collaborative way of learning. In this pedagogy, the student is at the center and follows teacher-guided principle. In this collaborative way of teaching learning process,

Fig. 2.2 Objectives of various pedagogical approaches

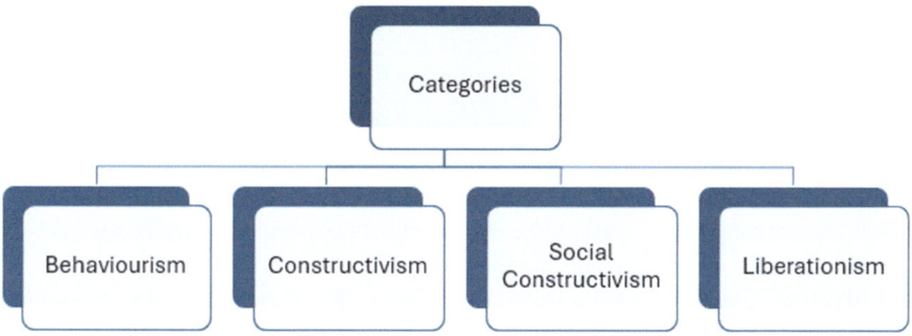

Fig. 2.3 Categories of pedagogical approaches

teacher is free to use mix mode of lessoning where discussion, modeling, questions can be the mode of teaching. It uses an integrated approach of social knowledge construction.

4. **Liberationism**

The democratic way of teaching in the classroom is the key principle of liberationism pedagogy. The subject under consideration is discovered together by the entire class and the teacher also plays the role of learner. Review of the state of the art, some game

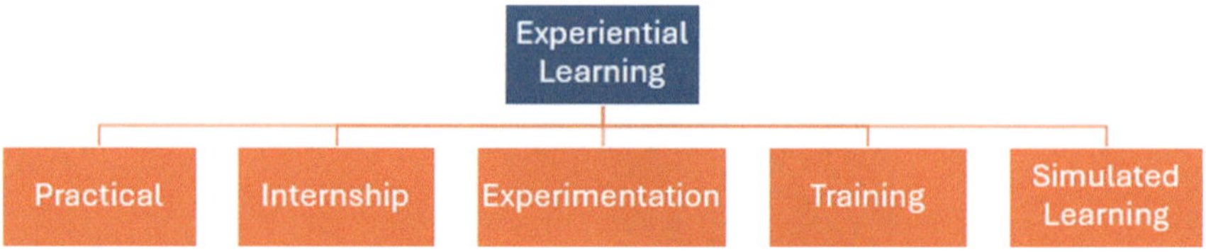

Fig. 2.4 Various tools used in experiential learning

playing activity, speech, dance are some activities adopted in liberationism pedagogy. Classroom democratization, thinking at critical level and learner independence are the founding principles of liberationism pedagogy.

4.1 Experiential Learning

Experiential learning enables learning by doing [6] where learner does not learn through various traditional pedagogies but learns through hands-on experiences. Experiential learning uses various tools like internships programs, research, practical, lab-based, and field-based experimentations, training, simulated training, etc. Experiential learning is most suited to learn new concepts in interdisciplinary fields where the learning curve is more because learners are empowered to take important decisions, ownership of the underlined problem as well as outcomes. Accountability on the outcomes by learner is more in experiential learning due to ownership. The level of ownership is more in experiential learning due to more critical thinking enabling more deep learning. Various tools used in experiential learning are depicted in Fig. 2.4.

The key benefits of experiential learning are listed below:

- **Personalization**: Personalization and personalized learning are the key benefit of experiential learning. Learning by doing also helps to create personal connection with the problem helping to improve learning curve. Personalized retention of knowledge and its application to solve practical problems through hands is also a main outcome of experiential learning.
- **Enhanced Creativity**: Creativity enhancement through hands on and practical exposure is another key benefit of experiential learning. This also helps learners to improve their novel and innovative skills.
- **Problem-Solving Skills**: Learning by doing helps learner to improve the problem solving and practical skill which is a prime need of corporate world. Different skill required for product building can be explored more effectively through experiential learning.
- **Collaborative Approach**: Experiential learning primarily helps learners to build their team work eventually fosters the collaborative approach of problem solving. This collaborative approach always helps learners to improve their communication skills and helps to create a more trusted ambience.
- **Research Rigor**: Enhanced research component is another important outcome of experiential learning. Strong mathematical foundations and critical thinking

developed through learning by doing plays an important role in developing the ability to bring novelty and innovation in the solution.

Experiential learning always works in the stages. The various stages of experiential learning are depicted in Fig. 2.5 and described below.

1. **Concept**: In experiential learning, the first and important step is to conceptualize the problem to be solved. In the sequel, it is very important to identify the valid problem to be solved.
2. **Cognizance**: The next step is to acquire detailed knowledge of the problem and understand every corner of the problem to be solved.
3. **Case**: Considering the use case or business case representing the underlined problem is the next step in the process of experiential learning. This sample case can later be explored for the purpose of systematic assessment.

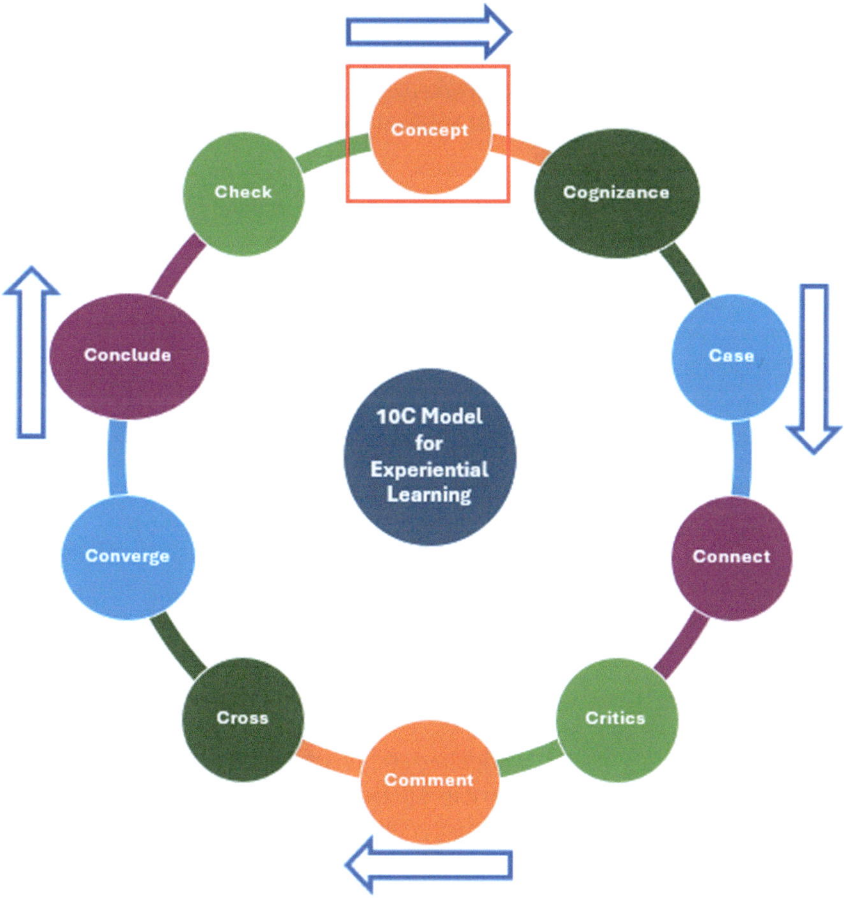

Fig. 2.5 10 C model for experiential learning

4. **Connect**: Connect with the above-mentioned use case to understand the set of questions to be posted and deriving the set of constraints to be associated with the solution.
5. **Critics**: Critical evaluation of the problem as well as the solution is a crucial step in experiential learning. Critical evaluation of the problem on both positive aspects and negative aspects always lead to a better solution.
6. **Comment**: Commenting the critical evaluation with detailed analysis helps to improve on the effectiveness of the solution. These comments can be incorporated and help to improve on the collaborative feature of experiential leaning.
7. **Cross**: The next step is to carry out cross examination of the solution for both functional and unit testing.
8. **Converge**: Converging solution in terms of upward and downward compatibility as well as interoperability plays a crucial role in experiential leaning to come up with the effective solutions.
9. **Conclude**: Concluding the process for checking out with some useful remarks so that it can be further reviewed and followed by the team as well as peer members is the next step in the experiential learning process.
10. **Check**: Finally checking out the solution for the purpose of deployment is the last step.

These are some recommended steps in experiential learning. However, these steps can be refined further for better understanding of the process and achieving optimal and useful solutions. It is all about how we learn naturally and apply it to problem solving. To summarize, in experiential learning learners take initiatives, make certain decisions and are accountable for the outcomes. IT provides opportunities toward learner to engage intellectually, creatively, emotionally, socially, or physically 2020 also put the same philosophy of introducing subject like robotics, coding, AI from the schooling to provide more hands-on experience to enable experiential learning in early days of the learner. This makes learning more relatable to the students and increases the effectiveness of learning by linking theory to practice. Increase in learner engagement, memory retention and leads to the development of skills for lifelong learning. The rest of chapter of this book talks about experiential learning for DM.

Moreover, every topic of the discrete mathematics will be learned through three stages

1. **Brainstorming Experiment (BE)**: The BE may be students' activities, some experiments, real case studies, etc.
2. **Inferences (I)**: The readers are advised to find out wonderful observations of the above experiment and write down in any language.
3. **Mathematical Insights (MI)**: Express the above inferences in the language of Mathematics that leads to a new mathematical theory.

These stages together, are known as the **"BEIMI"** philosophy or theory.

2.3 Discrete Mathematics in Engineering Applications

DM plays key roles in diversified engineering applications particularly in the design and modeling of the concepts and applications. Before the reader gets into learning it through experiential learning, it is important to explore the various tools of DM and their applications in the field. When we deal with the number, these can be continuous or discreet in nature. When the numbers are real in nature, it means that between any two continuous numbers in the series, there also exist continuous numbers. The example of continuous numbers is real numbers which satisfies the same property. However, the same property is not exhibited by discrete numbers. The example of discrete numbers is integer which do not exhibit the property mentioned above, i.e., between two integer there is not another integer present. In the sequel, DM is the study, exploration, and application of various mathematical objects which are discrete in nature. A few more examples of discrete objects include propositions in DM which is always either zero or one/ true or false/yes or no and there are not half-truth applicable in mathematics. Similarly, membership function for the set is always discrete, i.e., either an item is present in the set, or it is not present. There is no membership which is like partially in and partially out. Similar functionality is also applicable to the relations which makes it discrete. Graphs are another example of discrete component in DM where between every pair of vertices, there are no other vertices present.

Particularly referring to an example of computer science and engineering being most popular and interdisciplinary branch of engineering, there are many examples of the objects with discrete properties. Programming languages and their applications in multidisciplinary application development is increasing rapidly. Selectional programming construct in programming languages which includes if statement, if-else statements, and if-then-else statement are based on the propositional logic of DM. Digital logic also play a crucial role in electronic circuit design and is also based on the propositional logic of DM. Set theory, Boolean algebra is foundational concepts of DM which includes set creation, set operations and notations, set operations for building and reasoning of the sets. The set concepts of DM are heavily used in the hamming codes which is used in error correction and detection in the data storage as well as transmission.

Predicate logic plays an important role in defining preconditions, postconditions, spatial application development, formulation of temporal characteristics of the use case as well as mathematical modeling of the spatial and temporal systems. The field of security is becoming an integral part of every internet application where encryption and decryption are the key operations. For example, popular algorithm of RSA encryption is based on the DM concepts like mathematical reasoning, induction, and integer operations as well as discrete logarithmic problem. Functions and relations play a key role in the field of relational databases where all table relations and constraints are defined in the form of relations. Computer system modeling, their implementations, entity relationships, relational nomenclature, and specifications of computer-based systems are also based on the relations and functions concepts of DM.

Data structures like trees and graphs are also the core concepts of DM which has a wide range of applications in the field of computing [7]. Google maps, graph coloring problem, shortest path problem, spanning tress for computer, and communication networks, finding optimal binary search tress for message transmissions are the key applications. Power networks, data networks, network on the maps, PCB testing are based on these DM concepts of tress and graphs, and it is very important to make these concepts clear with practical applications.

Recurrence relations and counting principles are used in design and analysis of algorithms which is a core subject of computer science. For deriving the time complexity of any programming construct and algorithm, deriving recurrence relation, and then solving that recurrence relation to find the time complexity using asymptotic notations [8] are the key steps.

There are various applications of DM in computing where we deal with discrete objects. Consider the example of the entertainment industry which operates on multiple discrete objects and related operations on them. The video segment in the entertainment industry is based on the basic unit called pixels which are discrete objects in the domain. One segment of the video also consists of frames which are discreet in time. Color frequency of the video clip is also discretized into three main colors RGB, i.e., Red, Blue, and Green. The intensity factor for each color is also discrete in nature and it is 256 bits for every color.

Consider the real-world problem of academic activities at university level. In university, there are various courses which are being offered to the students at graduate and postgraduate level. Each course carries some credits and assessment of these courses includes in-semester examination as well as end semester examinations. This needs to assign some slots for examination to each course and the courses which are shared by one or more students should be assigned with different examinations slots. The problem is to decide on the assignment of slots to the courses and how to optimize the number of slots.

Consider the problem of coloring the nodes in graph with different colors so that no two adjacent nodes will have the same color. Similarly, there are other areas of engineering which include clustering and association mining problem, image analysis, and coloring the nodes or regions on the graphs with minimum colors so that no two adjacent regions will have the same colors. Programming languages also use graph coloring aspects when variables are assigned to the respective registers during the compilation of computer program, during the mapping of various functional units of hardware to various operations etc.

As presented and discussed in [7], there are six radio-frequency stations under surveillance to ensure that the signals generated by these radio-frequency stations don't interfere with each other. There is a central monitoring authority, i.e., Federal Communication Commission is responsible for this to assign appropriate frequencies to all the stations. The objective of this problem is to decide and confirm the minimum number of various frequency channels for these six stations so that any two stations can utilize the same frequency channel if these channels are located at more than a certain x distance.

2.4 Discrete Mathematics and Experiential Learning

Continuous mathematics and discrete mathematics are two key branches of mathematics which have their own sets of applications and boundaries. Quantity measurement is the main objective of continuous mathematics and finite decision making is key objective of discrete mathematics. Countable sets and uncountable sets are another two differentiating parameters between these two branches of mathematics respectively and the focus of discrete mathematics is always to determine the count for underlined dataset or set. The category of problems like optimization problems, counting problems, and existence problems are the three main problems which discrete mathematics deal with and are depicted in Fig. 2.6.

In the real-world every problem is an optimization problem where the objective is to maximize something or minimize something, i.e., finding an optimal solution for a given problem. Finding the shortest path from given source to given destination, maximizing sales of the goods to maximize the profit are some examples of optimization problems. Counting problem refers to the problem of utilizing given items in the space available. For example, how many different names of the person are possible with given number of characters. The last category of the problem in discrete mathematics is the existence problem which mainly deals with the verification of whether the solution exists or not. For example, in shortest path problems, the objective would be to first determine whether the path exists or not.

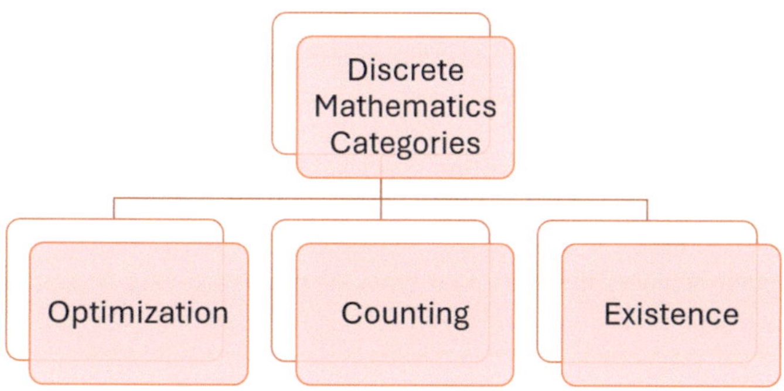

Fig. 2.6 Discrete mathematics problems

2.5 Applications

This section discusses applications of DM in the above-mentioned three categories of the problems which includes optimization, counting and existence problems.

Closure plays an important role in healthcare and engineering applications [9, 10]. In digitation and automation of healthcare and engineering applications, concepts of transitive and reflexive closure are vital. In computer and communication networks, the connectivity between computers and other computing nodes can be easily represented with the help of transitive and reflexive closures. Also, these closures are helpful in indicating the connectivity between two communication networks and their connectivity to in turn other computers and computing devices. This connectivity clearance also enables secure information exchange over the connected nodes. For example, the transitive and reflexive closure of this connectivity is mathematically represented as below:

$$\text{Is\_Connection} \rightarrow \text{Access\_Grant}$$

This relation is also helpful in achieving two fundamental principles of access control, i.e., principle of least privilege and selective disclosure [11]. Similar analogy is also used in healthcare during access control of electronic health records [12].

DM is very useful tool for design, analysis, and solution formulation of the problems which are combinatorial in nature. There are special categories of the problems where an attempt is to convert or have all or maximum variables in integer form, i.e., in discrete values and this special branch of DM is known as discrete optimization. Combinatorial optimization connects multiple disciplines of fundamental mathematics which includes mathematical and algorithmics complexity, operations research, linear programming and algorithmics. Routing problem in computer networks, bin packing problem, graph coloring, and graph covering problem are some examples of the combinatorial optimization problems. Various scheduling algorithms in the context of computer networks and operating systems as well as all sorting algorithms [8] are also the popular examples of this category. In the real world, there are many use cases where the solution exists subject to some constraints and these problems are referred to as constraint problems. Constraint programming is a useful programming technique to address and solve constraint programming [13, 14].

In computer science, there are two types of problems. One is a problem with output which generates some numeric, alpha-numeric or alphabetical output and another type of problem is a problem where the output is Boolean, i.e., output is either true or false/yes or no/0 or 1. The second category of the problem is referred to as decision problems [15]. Combinatorial decision problem [16] is a special type of decision problem where the objective is to find Boolean solution of the underlined problem with assurance of there exists optimal solution, i.e., finding a feasible solution for given criterion function. Permutation problems, Graph problems and finding optimal solution from countable sets are some popular examples of combinatorial optimization problems.

In the real world, DM plays vital role to solve counting problems especially in the form of combinatorics [17]. Finding the multiple possible combinations or arrangement of n numbers is the key objective of counting problems and DM is a popular tool for solving these counting problems. The rule of permutation and combination is used in many real-world use cases. For determining number of combinations in national id like Aadhaar number in Indian context or social security numbers for US citizens can be addressed using counting methods. Like this, there are many use cases where DM plays an important role in solving counting problems.

DM and existence problem [18] goes hand in hand. Various programming paradigm and design issues, programming constructs, software architectures, software design, theory of automata and cryptography are some examples and applications of DM which are design and developed using the theory of existence problem. In the field of computer and communication networks, design of communication protocols at all the layers as per OSI layers, routing algorithms, error correction, decoding, and alerting are also examples which are modeled using existence problem.

The rest of the book is organized as follows. Chapter 3 focuses on sets and elements, Venn diagrams, Set Operations, Algebra of Sets, Duality, Classes of Sets, Power Sets, Partitions, Inclusion and Exclusion of sets, and mathematical induction using experiential learning. Chapter 4 deals with the learning of relation using experiential learning. Chapter 5 explores the concepts of function which includes Functions and Partial Functions, Types of Functions, Infinite Sets and Countability, Pigeonhole Principle, Discrete Numeric Function and learning them with experiential Learning Chap. 6 deals with logic and proofs, Chap. 7 explores the concepts of permutations and combinations using experiential learning, Chap. 8 deals with discrete probabilities, Chap. 9 deals with graphs and Chap. 10 deals with trees. The concept of generating functions and recurrence relation and learning them using experiential learning is discussed in Chap. 11. Algebraic Structures with One Binary Operation and two binary operations using experiential learning is presented and discussed in Chaps. 12 and 13 respectively. Finally, Chap. 14 concludes the book with conclusions and future work.

2.6 Summary

For the teaching–learning process, the learning curve is an important parameter to measure outcome. In the emerging education industry as a service industry, outcome-based education is a very crucial measure for the evaluation. Experiential learning is one of the important learning methodologies to improve learning curve. There are various learning pedagogies in literature and experiential learning is found to be the better pedagogy for enhanced outcomes. This chapter mainly focuses on the different tools used in experiential learning and how DM can be effectively delivered to the learners. This chapter also discusses the unique 10C model proposed for experiential learning. The next part of the chapter discusses how DM is useful in engineering applications and how experiential learning can be useful in various categories of DM problems.

References

1. Rao J (2020) Outcome-based education: an outline. High Edu Futur 7:1–4
2. Jadhav MR, Kakade AB, Jagtap SR, Patil MS (2020) Impact assessment of outcome-based approach in engineering education in India. Procedia Comput Sci 172:791–796
3. Brahimi T, Sarirete A, Ibrahim R (2016) The impact of accreditation on student learning outcomes. Int J Knowl Soc Res 7(4):51–62
4. National education policy 2020. Ministry of Human Resource Development, Government of India
5. Salient features of NEP 2020: higher education. University Grants Commission, Bahadur Shah Zafar Marg, New Delhi
6. Kolb DA (2014) Experiential learning: experience as the source of learning and development. FT Press
7. Thadani S, Bagora S, Sharma A (2022) Applications of graph coloring in various fields. Mater Today Proc 66:3498–3501
8. Joshi Sarang A, Mahalle PN (2016) Design and analysis of algorithms: a problem-solving approach. India Cambridge University Press. ISBN: 9789385386527
9. Abo-Tabl EA, El-Bably MK (2022) Rough topological structure based on reflexivity with some applications. AIMS Math 7(6):9911–9999
10. Abu-Gdairi R, El-Gayar MA, Al-shami TM, Nawar AS, El-Bably MK (2022) Some topological approaches for generalized rough sets and their decision-making applications. Symmetry 14(1)
11. Mahalle PN, Anggorojati B, Prasad NR, Prasad R (2012) Identity driven capability-based access control (ICAC) for the internet of things. In: Proceedings of the 6th IEEE international conference on advanced networks and telecommunications systems (ANTS), Bangalore, India, IEEE, Dec 2012, pp 49–54
12. Mahalle PN, Patil RV, Dey N, Crespo RG, Sherratt RS et al (2023) Explainable AI for human-centric ethical IoT systems. IEEE Trans Comput Soc Syst 1–13. https://doi.org/10.1109/TCSS.2023.3330738
13. Wallace M (1996) Practical applications of constraint programming. Springer Constraints 1:139–168
14. Refalo P (2004) Impact-based search strategies for constraint programming. In: Principles and practice of constraint programming–CP 2004: 10th international conference, CP 2004, Toronto, Canada, Sept 27–Oct 1, 2004. Springer, Berlin, pp 557–571
15. Kalai A, Vempala S (2005) Efficient algorithms for online decision problems. J Comput Syst Sci 71(3):291–307
16. Post EL (1943) Formal reductions of the general combinatorial decision problem. Am J Math 65(2):197–215
17. Vilenkin NY (2014) Combinatorics. Academic Press
18. Aigner M (2023) Discrete mathematics. American Mathematical Society

Chapter 3
Sets

In mathematics, the art of asking questions is more valuable than solving problems—Georg Cantor

Abstract In real life, there are numerous situations where we need to collect the objects according to their characteristics or attributes and study those particular classes. This chapter deals with the concepts of sets, types of sets, operations, Venn diagram, algebra of sets, Inclusion-Exclusion Principles and Mathematical Induction. The chapter explains each concept by using the "BEIMI" philosophy. The example of the library system helps to enhance the understanding of the concepts of Inclusion and exclusion principles. It also explains how the cooking of any food is connected to the principle of Mathematical Induction. The theory is supported by precise real-life examples for meaningful understanding of the concepts. The chapter concludes with case studies, projects and exercises aimed at connecting theoretical relational structures to real-world applications across various disciplines.

Keywords Set · Power set · Inclusion-Exclusion principle · Mathematical induction principles

3.1 Introduction

In real life, there are numerous situations, where we need to collect the objects according to their characteristics or attributes and study that particular classes. The basic set theory developed by great mathematician Cantor and then many mathematicians contributed as per their necessity of the research [1, 2].

3.2 Sets, Elements, and Subsets: Experiential Learning

(A) **Brainstorming Experiment-I**

© The Author(s), under exclusive license to Springer Nature Singapore Pte Ltd. 2026

H. R. Bhapkar and P. N. Mahalle, *Discrete Mathematics: An Experiential Learning Perspective*, Studies in Autonomic, Data-driven and Industrial Computing,
https://doi.org/10.1007/978-981-95-7210-6_3

Let us consider the situation in the classroom. There are 30 boys and 20 girls in a class, and a teacher is teaching a subject of Discrete Mathematics. A teacher asked the students to find the collection of elements associated with the class. The students wrote the following collections.

1. The collection of boys in a class.
2. The collection of girls in a class.
3. The collection of all people present in a class.
4. The collection of honest students in a class.
5. The collection of beautiful girls in a class.
6. The collection of intelligent students in a class.

(B) **Inferences-I**

Let us consider the following notable inferences of the above activity.

1. In this collection, all boys are present. The collection is precisely defined. **[Well defined collection]**
2. All girl students are present in this collection. It is also defined precisely. **[Well defined collection]**
3. This collection involves a teacher, boys, and girls. It is well-defined collection. **[Well defined collection]**
4. The word "honest" is not defined precisely. So, we can't decide the collection precisely. But, we know that some or all boys could be in this collection. **[Not well defined collection]**
5. Similarly, the word "beautiful" is not defined properly. So, we can't find the collection. Some or all girls may or may not be in this collection.**[Not well defined collection]**
6. In this collection, some boys and girls are present. But we can't decide the exact collection. **[Not well defined collection]**

(C) **Mathematical Insights-I**

3.2.1 Sets

The set is the collection of well-defined objects. The objects can be anything, e.g., numbers, books, entity, article, etc. An object in the set H is recognized as an element of the set. In the sets, there are only two possibilities, whether an element is present in the set or not. So, the repetition of elements is not allowed in the set.

'$x \in H$' stands for "an element x is a member of the set H."
'$x \notin H$' stands for "an element x is not a member of the set H."

Examples of Sets:

1. Alphabets in the "Experiential Learning"
2. The books having the title "Discrete mathematics"

3. Passport numbers in the world.
4. All natural numbers.
5. All students who opted the course of Discrete Mathematics.

Examples which are not Sets:
The following are the examples which are not sets.

1. The collection of all clever students in the class
2. The collection of all beautiful flowers in the garden
3. The collection of all honest citizens in a particular district.
4. The collection of good books in the library.
5. The collection of ethical persons in the society.

The sets are described by three different methods.

i. **Roster Method (Listing Method)**
In this method, the elements of the set are listed between a pair of brackets/ braces. e.g., $H = \{1, 2, 3, 4, 5\}$, $K = \{a, c, e, g, h\}$.
ii. **Statement Form**
If the sets are formed by using the statements or common attributes of the elements, then it is known as the statement form of the set, e.g., the set of all even integers, the set of all alphabets in English language, the set of all students from a particular district.
iii. **Set Builder Notation**
It is a more compact and concise way of describing the sets.
e.g., $H = \{h/h \text{ is an integer and } h \geq 13\}$ [3, 4]

The following are some important sets of the numbers, which helps us to understand the mathematical theory more precisely.

S. No.	Name of the set	Notation and elements
1	Natural numbers set or positive integers set	$N = \{1, 2, 3, \ldots\}$
2	Whole numbers set or non-negative integers set	$W = \{0, 1, 2, 3, \ldots\}$
3	Integers set	$Z = \{\ldots, -3, -2, -1, 0, 1, 2, 3, \ldots\}$
4	Rational numbers set	$Q = \left\{\frac{h}{m}/h, m \in Z, (h, m) = 1, m \neq 0\right\}$
5	Real numbers set	R
6	Complex numbers set	$C = \{h + im/h, m \in R, i = \sqrt{-1}\}$
7	Non-zero real numbers set	$R_0 = R - \{0\}$

3.2.2 *Subsets*

A set M is said to be subset of the set H if $x \in M \Rightarrow x \in H, \forall x \in M$.

It is denoted by $M \subseteq H$. If $M \subseteq H$ then H is known as the superset of M. The null set ϕ or $\{\}$ is empty set [5].

Depending upon the nature of the elements in the subset, there are two types of subsets.

1. **Proper Subsets**: A subset M is called a proper subset of H if $\exists h \in H$ but $h \notin M$.
2. **Improper Subsets**: The subsets ϕ and H are known as the improper subsets of the set H.

Note:

If the set H is formed with respect to the property P, then it's all elements must follow the same property P. But, the other properties may or may not be the same. For example, suppose H and M are integers set and positive even integers set, respectively. Now, the set having only positive integers is a subset of H, not M.

Question for Readers: Whether the subsets of the subsets of the set H, are the subsets of H. Justify it.

3.3 Venn Diagram: Experiential Learning

The pictorial representations of mathematical terms or theory are easy to understand and apply it to resolve real-life issues. The pictorial representation of sets by using a particular set of diagrams is used by British Mathematician John Venn. The pictorial representation of sets which is designed by using the following rules is called as the Venn diagram [5, 6].

1. The Universal Set U is represented by a big rectangle/circles.
2. The subsets of U are represented by circles or some closed curve.
3. If $M \subset H$ then the circle of M lies inside the circle of H.
4. If M and H are disjoint sets then their circles do not have common area.
5. If $M \cap H \neq \phi$ then the circles of M and H have some common area.

Figure 3.1 depicts the set and subsets.

3.4 Set Operations: Experiential Learning

We know that in real-number system, new numbers are generated by using four basic operations. Similarly, the new sets are formed by using the union, intersection, complement, and difference of sets [1, 7].

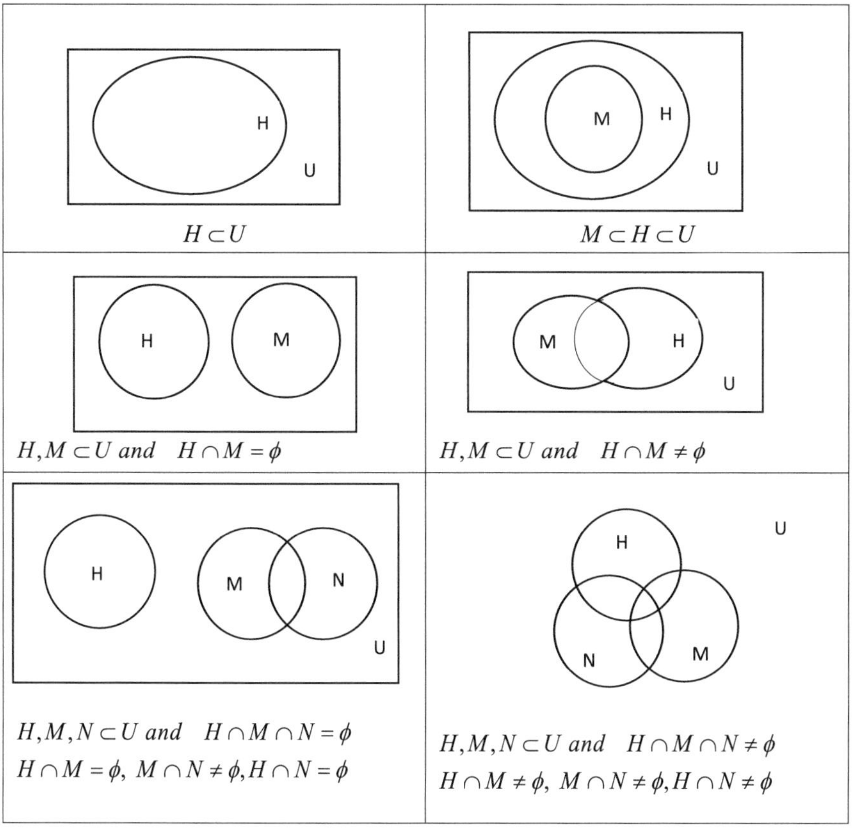

Fig. 3.1 Representation sets and subsets by Venn diagram

3.4.1 Intersection of Two Sets

The intersection of two sets H and M (denoted by $H \cap M$) defined as $H \cap M = \{x/x \in H \text{ and } x \in M\}$. There are two possibilities either $H \cap M = \phi$ or $H \cap M \neq \phi$. It's Venn diagram is as follows shown in Fig. 3.2.

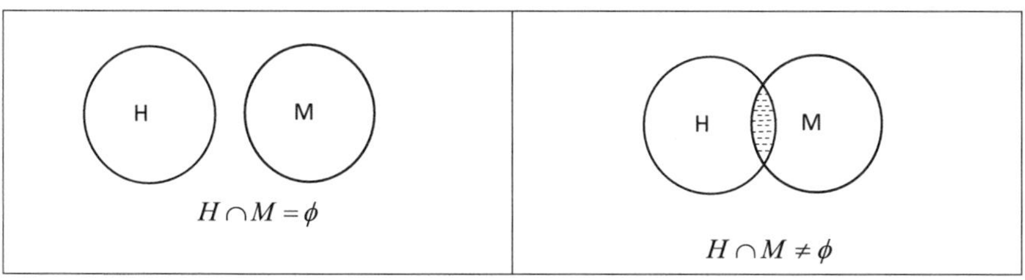

Fig. 3.2 Representation intersection of two sets

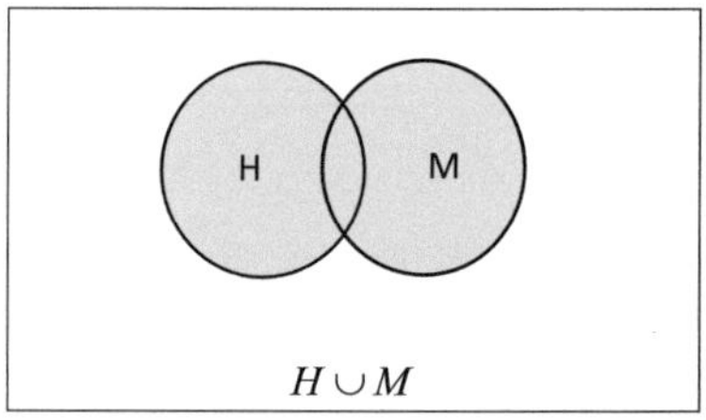

Fig. 3.3 Representation union of two sets

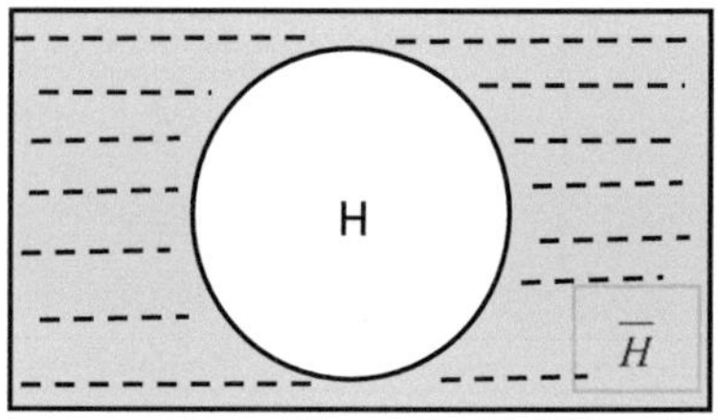

Fig. 3.4 Representation complement of a set

Note: Two sets H and M are said to be disjoints if $H \cap M = \phi$.

3.4.2 Union of Two Sets

The union of two sets H and M is defined as $H \cup M = \{x/x \in H \text{ or } x \in M \text{ or } x \in H \cap M\}$.

By Venn diagram, the $H \cup M$ is depicted Fig. 3.3.

3.4.3 Complement of a Set

The complement of a set H is defined as $\overline{H} = \{x/x \notin H\}$. By Venn diagram, the $\overline{H}$ is represented in Fig. 3.4.

Properties of Complement of a Set

1. $\overline{U} = \phi$ *and* $\overline{\phi} = U$
2. $\overline{(\overline{H})} = H$
3. $I \cup \overline{I} = U$ *and* $I \cap \overline{I} = \phi$.

3.4.4 Difference of Two Sets

The difference of H and M is defined as $H - M = \{h/h \in H \text{ and } h \notin M\}$. It is also known as the relative complement of M in H.

Fig. 3.5 Representation of $H - M$

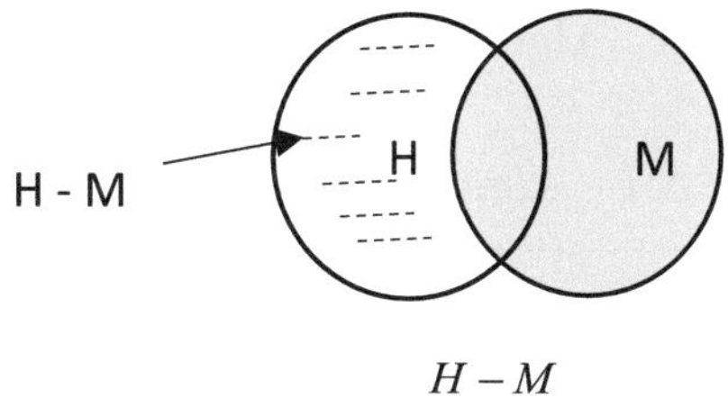

By Venn diagram, the $H - M$ is represented as follows, depicted in Fig. 3.5.

Properties:

Let H and M be any two sets. Then

1. $\overline{H} = U - H$
2. $I - I = \phi$ and $\phi - \phi = \phi$
3. $I - \overline{I} = I, \overline{I} - I = \overline{I}$ and $U - \overline{\phi} = \phi$
4. $I - \phi = I$

3.4.5 Symmetric Difference of Two Sets

The symmetric difference of sets H and M is defined as

$H \oplus M = \{\alpha/\alpha \in H - M$ or $\alpha \in M - H\}$. By Venn diagram, the $H \oplus M$ is represented as follows (Fig. 3.6).

Properties of $H \oplus M$

Let I and W be any two sets. Then

1. $I \oplus I = \phi$
2. $I \oplus \phi = I$
3. $I \oplus U = U - I = \overline{I}$
4. $I \oplus \overline{I} = U$
5. $I \oplus W = (I \cup W) - (I \cap W)$

Fig. 3.6 Representation symmetric difference of two sets

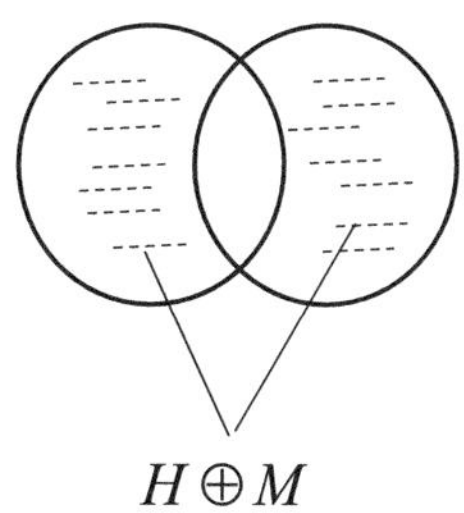

3.5 Algebra of Sets and Duality: Experiential Learning

3.5.1 *Algebra of Sets*

The set operations follow the similar rules as that of the rule in number theory. Let us consider the following algebra of the set of operations [2, 8].

Theorem 3.1 *Let* I, M, S *be any three sets. Then*

1. *Commutative Laws with respect to* $\cup$ *and* $\cap$
   (a) $I \cup M = M \cup I$ (b) $I \cap M = M \cap I$
2. *Associative Laws with respect to* $\cup$ *and* $\cap$
   (a) $I \cup (M \cup S) = (I \cup M) \cup S$ (b) $I \cap (M \cap S) = (I \cap M) \cap S$
3. *Distributive Laws*
   (a) $I \cup (M \cap S) = (I \cup M) \cap (I \cup S)$
   (b) $I \cap (M \cup S) = (I \cap M) \cup (I \cap S)$
4. *Idempotent Laws*
   (a) $I \cup I = I$ (b) $I \cap I = I$
5. *Absorption Laws*
   (a) $I \cup (I \cap M) = I$ (b) $I \cap (I \cup M) = I$
6. *De Morgan's Laws*
   (a) $\overline{(I \cup M)} = \overline{I} \cap \overline{M}$ (b) $\overline{(I \cap M)} = \overline{I} \cup \overline{M}$
7. *Double Complement:* $\overline{(\overline{I})} = I$

Proof These proofs are easy, so readers can prove it. Here, we will prove (3) and (5) only.

3. Prove that $I \cup (M \cap S) = (I \cup M) \cap (I \cup S)$

$$\begin{aligned} \text{Let} x \in I \cup (M \cap S) &\Rightarrow x \in I \text{ or } x \in M \cap S \\ &\Rightarrow x \in I \text{ or } (x \in M \quad \text{and } x \in S) \\ &\Rightarrow (x \in I \text{ or } x \in M) \text{ and } (x \in I \text{ or } x \in S) \\ &\Rightarrow (x \in I \cup M) \text{ and } (x \in I \cup S) \\ &\Rightarrow x \in (I \cup M) \cap (I \cup S) \end{aligned}$$

Therefore, $I \cup (M \cap S) \subseteq (I \cup M) \cap (I \cup S)$.
Similarly, we can prove that $I \cup (M \cap S) \supseteq (I \cup M) \cap (I \cup S)$. Hence the proof.

5. Prove that $I \cup (I \cap M) = I$.

$$\text{Let} x \in I \cup (I \cap M) \Rightarrow x \in I \text{ or } x \in (I \cap M)$$

$$\Rightarrow x \in I \text{ or } (x \in I \text{ and } x \in M)$$
$$\Rightarrow (x \in I \text{ or } x \in I) \text{ and } (x \in I \text{ or } x \in M)$$
$$\Rightarrow (x \in I) \text{ and } (x \in I \cup M)$$
$$\Rightarrow x \in I$$

Therefore, $I \cup (I \cap M) \subseteq I$. Similarly, $I \subseteq I \cup (I \cap M)$. Hence the proof.

3.5.2 Duality of Sets

The Principle of duality states that "Any proved results/theorems involving sets, complements and operations of Union and Intersection gives a corresponding dual results/theorems by replacing U by ϕ, $\cup$ by $\cap$ and vice versa".

The dual of $H \cup \overline{H} = U$ is $H \cap \overline{H} = \phi$.

3.6 Classes of Sets, Power Set, and Partitions: Experiential Learning

Depending upon some attributes/ properties, there are following main types of sets.

1. **Singleton Set**: It has exactly one element, e.g., $H = \{13\}$
2. **Finite Set**: A set H is called a finite set if the number of elements in H is finite, i.e., $|H| =$ finite.
3. **Infinite Set**: A set H is an infinite set if $|H| = \infty$.
4. **Power Set**: The power set of the set H is defined as
   $P(H) = \{X/X \subseteq H\}$
   $P(H) =$ the set of all subsets of the set H.

Set (H)	Power set $P(H)$
ϕ	$\{\phi\}$
$\{x\}$	$\{\phi, \{x\}\}$
$\{h, p\}$	$\{\phi, \{h\}, \{p\}, \{h, p\}\}$
$\{h, p, z\}$	$\{\phi, \{h\}, \{p\}, \{z\}, \{h, p\}, \{p, z\}, \{h, z\}, \{h, p, z\}\}$

Note: If $|H| = n$ then $|P(H)| = 2^n$ and $|P(P(H))| = 2^{2^n}$.

3.7 Inclusion and Exclusion of Sets: Experiential Learning

(A) Brainstorming Experiment-II

Let H and M be the set of students, who like Mathematics and Data Structure, respectively. We assume that if a student got more than 70% marks in a subject, in the end-of-semester examination, then the student likes that subject. There are 35 students in the set H, and 25 students in the set M. Let us consider the following possible cases.

1. The set $H \cap M$ is empty.
2. The set $H \cap M$ is non-empty.

(B) Inferences-II

The following are the notable inferences of the above experiment.

1. As $H \cap M$ is an empty set, and $|H \cup M| = |H| + |M|$
2. If the set $H \cap M$ is a non-empty, and $|H \cup M| < |H| + |M|$

In this case, the main question is how to find the number of elements in $H \cup M$, suppose there are five students in both H and M. So, to find the number of students in $H \cup M$, add the students in H, and the students in M, and to avoid repetition, delete the students that are in $H \cap M$. Therefore, we find the number of students in $H \cup M$ by including and excluding the students from the sets. Thus, this method is known as the method of Inclusion and Exclusion.

(C) Mathematical Insights-II

Cardinality of a Set:
The cardinality of the set H is $|H| = n(H) =$ Number of elements in H.

Note: The cardinality of the null set is zero, i.e., $n(\phi) = 0$.

Theorem 3.1 *If I and W be two finite sets and $I \cap W = \phi$ then $|I \cup W| = |I| + |W|$.*

Proof The proof is easy, so the readers can prove it.

Theorem 2: If $I_1, I_2, I_3, \ldots, I_n$ be mutually disjoint finite sets then

$$|I_1 \cup I_2 \cup I_3 \cup \cdots \cup I_n| = |I_1| + |I_2| + |I_3| + \cdots + |I_n|.$$

Theorem 3.2 *If S is a finite set and M is any set then $|S - M| = |S| - |S \cap M|$.*

Proof By the Venn diagram depicted in Fig. 3.7

$$S = (S - M) \cup (S \cap M).$$

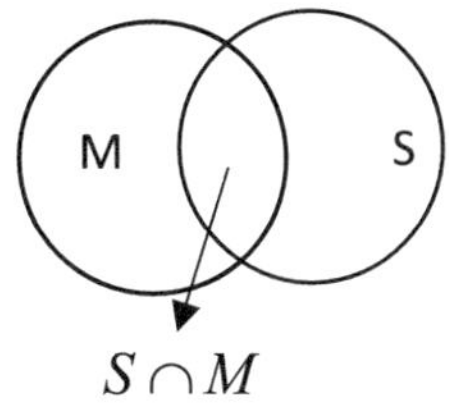

Fig. 3.7 Intersection of two sets

The sets $(S - M)$ and $(S \cap M)$ are disjoint and finite sets. Therefore, by Theorem 1,

$$|S| = |S - M| + |S \cap M| \quad \Rightarrow \; |S - M| = |S| - |S \cap M|.$$

The Principle of Inclusion and Exclusion for Sets

Theorem 3.3 *The Principle of Inclusion and Exclusion for two sets:*
If S and M be any two finite sets, then $|S \cup M| = |S| + |M| - |S \cap M|$.

Proof By the Venn diagram, the sets $S - M$ and M are disjoint and finite sets. Therefore, by Theorem 1,

$$|S \cup M| = |S - M| + |M| = |S| - |S \cap M| + |M| = |S| + |M| - |S \cap M|.$$

Theorem 3.4 *The Principle of Inclusion and Exclusion for three sets:*
If S, M, and P be any three finite sets then

$$|S \cup M \cup P| = |S| + |M| + |P| - |S \cap M| - |S \cap P| - |M \cap P| + |S \cap M \cap P|.$$

Proof Let $X = M \cup P$. Therefore, H and X are two finite sets, then

$$|S \cup X| = |S| + |X| - |S \cap X|.$$

Therefore,

$$\begin{aligned}|S \cup M \cup P| &= |S \cup X| = |S| + |X| - |S \cap X| \\ &= |S| + |M \cup P| - |S \cap (M \cup P)| \\ &= |S| + |M| + |P| - |M \cap P| - |(S \cap M) \cup (S \cap P)| \\ &= |S| + |M| + |P| - |M \cap P| - |S \cap M| - |S \cap P| + |S \cap M \cap P| \\ &= |S| + |M| + |P| - |S \cap M| - |S \cap P| - |M \cap P| + |S \cap M \cap P|\end{aligned}$$

Hence the proof.

Similarly, we can apply this result for the finite number of sets.

Example 3.1 In a survey of 100 students from academic institution, it is observed that 32 students study Discrete Mathematics (DM), 20 students study Python language (PL), 45 students study Data Structure (DS), 15 students study Discrete Mathematics and Data Structure, 07 students study Discrete Mathematics and Python language, 10 students study Python (PL) and Data Structure, and 30 students do not study any of the three subjects.

(a) Find the number of students studying DM, PL, and DS
(b) Find the number of students studying exactly one of the 3subjects.
(c) Find the number of students studying exactly two of the 3 subjects.
(d) Find the number of students studying exactly DM and PL.
(e) Find the number of students studying DM but not PL.

Solution Let *S*, *M*, *P* be the set of students studying DM, PL, and DS, respectively. Therefore,
$|U| = 100, |S| = 32, |M| = 20, |P| = 45, |S \cap M| = 7, |S \cap P| = 15, |M \cap P| = 10$, and $\left|\overline{S} \cap \overline{M} \cap \overline{P}\right| = 30$.

$$\therefore \left|\overline{S} \cap \overline{M} \cap \overline{P}\right| = 100 - |S \cup M \cup P|$$
$$\Rightarrow |S \cup M \cup P| = 100 - 30 = 70$$

(a) The number of students studying all three subjects is obtained as follows.

$$|S \cup M \cup P| = |S| + |M| + |P| - |S \cap M| - |S \cap P| - |M \cap P| + |S \cap M \cap P|$$
$$\Rightarrow 70 = 32 + 20 + 45 - 7 - 15 - 10 + |S \cap M \cap P|$$
$$\Rightarrow |S \cap M \cap P| = 70 - 65 = 5$$

Thus, 05 students are studying all three subjects.

(b) The number of students studying exactly one subject:
The number of students studying only DM is

$$|S| - |S \cap M| - |S \cap P| + |S \cap M \cap P| = 15.$$

The number of students studying only PL is

$$|M| - |M \cap S| - |M \cap P| + |S \cap M \cap P| = 8.$$

The number of students studying only DS is

$$|P| - |P \cap S| - |P \cap M| + |S \cap M \cap P| = 45 - 15 - 10 + 5 = 25.$$

Therefore, the number of students studying exactly one subject is $= 15 + 8 + 25 = 48$.

(c) The number of students studying exactly two subjects is

$$|S \cap M| + |S \cap P| + |M \cap P| - |S \cap M \cap P| = 7 + 15 + 10 - 5 = 27.$$

(d) The number of students studying exactly DM and PL is

$$|S \cap M| - |S \cap M \cap P| = 7 - 5 = 2.$$

(e) The number of students studying DM but not PL is

$$|S - M| = |S| - |\cap M| = 32 - 7 = 25.$$

3.8 Mathematical Induction: Experiential Learning

(A) **Brainstorming Experiment-III**

Let us try to find the answer to the common question, how to check whether the rice is cooked properly or not. We check 2–5 grains of rice and if these are cooked properly then, we assume that all the rice is cooked properly.

B) **Inferences-III**

Consider the following significant observations.

1. All grains of rice have the same property.
2. If a few grains of the rice change their property, then we assume that all grains of the rice change their property.

(C) **Mathematical Insights-III**

The positive integers set has a special importance in the discrete Mathematics for proving particular types of results. The set of positive integers and the set of natural numbers are equal and denoted as $N = Z^+ = \{1, 2, 3, \ldots\}$.

The Well-Ordering Principle: Let $H \subseteq N = Z^+$ and $H \neq \phi$ then H has the smallest element, i.e., $\exists s \in H$, such that $s \leq x, \forall x \in H$ [5].

There are two types of the principle of Mathematical Induction.

i. First Principle of Mathematical Induction
ii. Second Principle of Mathematical Induction.

3.8.1 First Principle of Mathematical Induction

Let $H(w)$ be a statement involving a natural number $w \geq w_0$such that

1. *Basis of Induction: If $H(w_0)$ is true, and*

2. *Induction Step: If whenever, $H(p)$ is true then $H(p+1)$ is also true;*

Then, $H(w)$ is true for all $w \geq w_0$.

Proof Let $H(w)$ be a statement involving a natural numbers $w \geq w_0$ satisfying conditions 1 and 2. Let $M = \{x \in N / H(x) \text{ is false}\}$.

Claim: Prove that $M = \phi$.

Assume that $M \neq \phi$. By well-ordering principle, the set M has a least element say s. As $H(w_0)$ is true. Therefore, $s \neq w_0$ and $s > w_0$, so, $(s-1) \in N$. But $(s-1) \notin M$. We have $H(s-1)$ is true. Therefore, by the induction step, $H(s-1+1) = H(s)$ is true. This contradiction to $s \in M$. Therefore, $M = \phi$. Hence the proof.

3.8.2 Second Principle of Mathematical Induction

Let $H(w)$ be a statement involving a natural number $w \geq w_0$such that

1. *Basis of Induction: If $H(w_0)$ is true, and*
2. *Induction Step: Assume that $H(w)$ is true for $w_0 \leq w \leq p$, i.e., $H(w_0), H(w_0+1), H(w_0+2), \ldots, H(p)$ are true then $H(p+1)$ is true;*
   Then, $H(w)$ is true for all $w \geq w_0$.

Example 3.2 Prove that $1^3 + 2^3 + 3^3 + \cdots + w^3 = \frac{w^2(w+1)^2}{4}$ where $w \in N$.

Solution Let $H(w)$ be the given statement.

1. Basis of Induction:
   For, $n = 1$, LHS $= 1$ and RHS $= \frac{1(1+1)^2}{4} = 1 \Rightarrow$ LHS $=$ RHS.
   Hence, $H(1)$ is true.
2. Induction Step: Assume that $H(p)$ is true.
   i.e., $1^3 + 2^3 + 3^3 + \cdots + p^3 = \frac{p^2(p+1)^2}{4}$. Then, we have

$$1^3 + 2^3 + 3^3 + \cdots + p^3 + (p+1)^3 = \frac{p^2(p+1)^2}{4} + (p+1)^3$$
$$= (p+1)^2\left\{\frac{(p+2)^2}{4}\right\} = \frac{(p+1)^2(p+2)^2}{4}$$

Therefore, $H(p+1)$ is true. Thus, $H(w)$ is true, for all w.

Example 3.2 Prove that $8^n - 3^n$ is a multiple of 5 for $n \geq 1$.

Solution Let $H(w)$ be the given statement, i.e., $H(w) : 8^w - 3^w = 5r, r \in N$.

1. Basis of Induction:
   For, $w = 1, 8^1 - 3^1 = 5 = 5 \times 1$. Hence, $H(1)$ is true.

2. Induction Step: Assume that $H(p)$ is true.
   i.e., $H(p) : 8^p - 3^p = 5r, r \in N$. Then, we have

$$\begin{aligned}8^{p+1} - 3^{p+1} &= 8^k \times 8 - 3^p \times 3 = 8^p \times (5+3) - 3^p \times 3 \\ &= 8^p \times 5 + 8^p \times 3 - 3^p \times 3 \\ &= 8^p \times 5 + (8^p - 3^p) \times 3 \\ &= 8^p \times 5 + (5 \times r) \times 3 \\ &= (8^p + r \times 3) \times 5\end{aligned}$$

   Therefore, $H(p+1)$ is true. Thus, $H(w)$ is true for all $w \geq 1$.

Example 3.3 Prove that $w \geq 14$ can be written as the sum of 3's and/or 8's.

Solution Let $H(w)$ be the given statement.

1. Basis of Induction:
   For, $w = 14$, $14 = 8 + 3 + 3$. Hence, $H(14)$ is true.
2. Induction Step: Assume that $H(p)$ is true for $14 \leq w \leq p$, i.e., $H(14), H(15), H(16), \ldots, H(p)$ are true.
   Now, consider

$$\begin{aligned}p + 1 &= (p-2) + 3 \\ &= 3a + 8b + 3; \quad \text{for some } a \text{ and } b \\ &= 3(a+1) + 8b\end{aligned}$$

   Therefore, $H(p+1)$ is true. Thus, $H(w)$ is true for all $w \geq 14$.

3.9 Case Studies and Projects

Consider the following case studies/projects.

1. Suppose there are four main political parties contested a particular election in your State. Represent the vote distributions of the parties by set theory. Represent the vote distribution of your district by Venn diagram.
2. Represent the result analysis of the end semester examination of your class/ college by Venn diagram.
3. Out of 60 students in a college, 30 take an A course, 16 take M course, 12 take P course, and 6 students take all courses. Show that 14 or more students take none of the courses. Use Venn diagram and inclusion exclusion theorem.
4. Find out the survey of all professors in your college about their liking of sports. Take three options from each. Do analysis of it by designing your own questions.
5. Write report on at least five applications of principle of Mathematical Induction.

Exercise

1. Prove that the empty set is the subset of every set.
2. Let H, M, P be any three sets. Then

   i. Commutative Laws with respect to $\cup$ and $\cap$
   ii. Associative Laws with respect to $\cup$ and $\cap$
   iii. Distributive Laws
   iv. Idempotent Laws
   v. Absorption Laws
   vi. De Morgan's Laws
   vii. Double Complement:

3. Prove the following by Venn diagram.

   (a) $H \cap (M \oplus N) = (H \cap M) \oplus (H \cap N)$
   (b) $(H - M) \cup (H - N) = H$

4. Consider the set of integers from 1 to 1000. Find

   (a) How many of these integers are divisible by 3 or 5 or 11?
   (b) How many of these integers are divisible by 3 or 5 but not by 11?
   (c) How many of these integers are divisible by 3, but not by 5 or 11?
   (d) How many of these integers are not divisible by 3, and 5, and 11?

5. Prove that

   (a) $1 + 2 + 3 + \cdots + w = \frac{w(w+1)}{2}$,
   (b) $\frac{1}{1\times2} + \frac{1}{2\times3} + \frac{1}{3\times4} + \cdots + \frac{1}{w\times(w+1)} = \frac{w}{w+1}$,
   (c) $\frac{1}{1\times3} + \frac{1}{3\times5} + \frac{1}{5\times7} + \cdots + \frac{1}{(2w-1)\times(2w+1)} = \frac{w}{2w+1}$.

6. Show that the sum of the cubes of the three consecutive natural numbers is divisible by 9.

References

1. Aigner M (2023) Discrete mathematics. American Mathematical Society
2. Erciyes K (2021) Discrete mathematics and graph theory. Springer, Cham. https://doi.org/10.1007/978-3-030-61115-6
3. Rosen K, Krithivasan K (2021) Discrete mathematics and its applications. Mc Graw Hill India. ISBN-978930727353
4. Jiri M, Jaroslav N (2011) Invitation to discrete mathematics, 2nd edn. Oxford University Press. ISBN-9780198570424
5. Peter J (2000) Sets, logic and categories. Springer Publication. ISBN-9781852330569
6. Tremblay JP, Manohar R (2023) Discrete mathematical structures with applications to computer science. McGraw Hill Education. ISBN-9780074631133
7. Vilenkin NY (2014) Combinatorics. Academic Press
8. Koshy T (2004) Discrete mathematics with applications. Elsevier Academic Press, Burlington, MA

Chapter 4
Relations

Mathematicians investigate relations among objects rather than objects themselves; they don't care if objects are replaced with other objects as long as the relationships remain the same. They are just interested in the forms; matter is unimportant—Henri Poincaré (Mathematician)

Abstract The chapter 4 provides an inclusive study of the mathematical theory of Relations, transitioning from the fundamental concepts of set theory to their practical and abstract applications in Discrete Mathematics. It begins by introducing the Cartesian Product of sets as the structural foundation for defining relationships between elements of objects of sets. Through a series of "Experiential Learning" experiments and brainstorming sessions, ranging from student college admissions to social dynamics, the text illustrates how complex real-world scenarios are modeled using mathematical notations. The chapter systematically categorizes various types of relations, including reflexive, symmetric, transitive, antisymmetric, and compatible relations, while providing rigorous definitions and proofs for each. Advanced topics such as composite relations, equivalence classes, set partitions, and Partial Ordering Relations (POSETS) are also discussed. It also focuses on the closure properties, offering step-by-step algorithmic approaches like Warshall's Algorithm and Matrix Method for determining transitive closures. The chapter concludes with case studies, projects and exercises designed to bridge the gap between theoretical relational structures and their multidisciplinary applications.

Keywords Cartesian product · Digraphs · Equivalence relation · POSET · Transitive closure · Warshall's algorithm

4.1 Introduction

In the previous chapter, we dealt with the sets, their types, properties, and examples. There could be various relationships among elements of the sets with respect to some desirable parameters or attributes. Actually, the name "Relation" itself suggests the

© The Author(s), under exclusive license to Springer Nature Singapore Pte Ltd. 2026

H. R. Bhapkar and P. N. Mahalle, *Discrete Mathematics: An Experiential Learning Perspective*, Studies in Autonomic, Data-driven and Industrial Computing,
https://doi.org/10.1007/978-981-95-7210-6_4

intended meaning. In the dictionary, the definition is that the relation is the connection between two or more things with respect to some properties or characteristics [1].

In general, our Mother, Father, Brother, Sister, Friend, Teacher, and Mentor are examples of the relationship among you and your family members or others. Besides these relations, we could compare different cities, villages, buildings, rivers, books, etc., concerning comparable attributes. Two buildings are identical or equivalent with respect to their heights.

Two books are identical concerning to their price. But, if we compare the quality of two books of the different categories by making the survey from the same group of people, we can find the biased results on quality of books depending upon the majority. Moreover, if the one book is on "Cooking of Delicious Foods" and the other is on "Designing Innovative Thinking." People who are very foodie surely will give the first preference to the book on food and others will to the book of the Innovative thinking. These two objects are from the same family, i.e., books, but these two are not comparable as the word quality is subjective. In this context, we should take two books from the same subcategory and define the concrete criteria for the quality aspects. This would help us to decide the ranking of books with respect to desirable attributes or parameters [2, 3].

4.2 Product of Sets: Experiential Learning

Let's consider the following brainstorming Experiment-I for understanding the concept of the product of the sets.

(A) **Brainstorming Experiment-I**

There is a group of ten students who have recently passed 12th standard examination in Science and want to get admission for Bachelor of Technology program. The students have collectively identified the six best colleges/Institutions as per their requirements. So, they want to get admission to these institutions only.

Let S_1, S_2, S_3, …, S_{10} be the name of ten students and C_1, C_2, C_3, …, C_6 be the name of six colleges, respectively. Let us find all possible combinations of the experiment. Every student has 6 possibilities or options for getting admission for B.Tech. Program. The following set S denotes the all possible combinations.

$$S = \left\{ \begin{array}{c} (S_1, C_1), (S_1, C_2), (S_1, C_3), (S_1, C_4), (S_1, C_5), (S_1, C_6), \\ (S_2, C_1), (S_2, C_2), (S_2, C_3), (S_2, C_4), (S_2, C_5), (S_2, C_6), \\ (S_3, C_1), (S_3, C_2), (S_3, C_3), (S_3, C_4), (S_3, C_5), (S_3, C_6), \\ (S_4, C_1), (S_4, C_2), (S_4, C_3), (S_4, C_4), (S_4, C_5), (S_4, C_6), \\ (S_5, C_1), (S_5, C_2), (S_5, C_3), (S_5, C_4), (S_5, C_5), (S_5, C_6), \\ (S_6, C_1), (S_6, C_2), (S_6, C_3), (S_6, C_4), (S_6, C_5), (S_6, C_6), \\ (S_7, C_1), (S_7, C_2), (S_7, C_3), (S_7, C_4), (S_7, C_5), (S_7, C_6), \\ (S_8, C_1), (S_8, C_2), (S_8, C_3), (S_8, C_4), (S_8, C_5), (S_8, C_6), \\ (S_9, C_1), (S_9, C_2), (S_9, C_3), (S_9, C_4), (S_9, C_5), (S_9, C_6), \\ (S_{10}, C_1), (S_{10}, C_2), (S_{10}, C_3), (S_{10}, C_4), (S_{10}, C_5), (S_{10}, C_6), \end{array} \right\} \tag{4.1}$$

(B) **Inferences-I**

Here, we have two sets: (1) Set of Students and (2) Set of Colleges/Institutions. Every set is non-empty set. There are $10 * 6 = 60$ elements in the set S that includes all possibilities. The first element of the order pair is the student and the second is the College. The set of all possibilities is expressed as follows.

(C) **Mathematical Insights-I**

Cartesian Product of Sets

The Cartesian Product (CP) of two sets H and P is defined as

$$H \times P = \{(u, w)/u \in H \text{ and } w \in P\}$$

For Example: If $H = \{11, 12, 13, 14\}$ and $P = \{x, y, w\}$ then

$$H \times P = \left\{ \begin{array}{l} (11, x), (11, y), (11, w), \\ (12, x), (12, y), (12, w), \\ (13, x), (13, y), (13, w), \\ (14, x), (14, y), (14, w) \end{array} \right\}$$

$$P \times H = \{(x, 11), (x, 12), (x, 13), (x, 14), (y, 11), (y, 12), (y, 13), (y, 14), (w, 11), (w, 12), (w, 13), (w, 14)\}$$

Here, $(11, x) \neq (x, 11)$. Hence, $H \times P \neq P \times H$. The Cartesian product of two sets is not commutative.

In general, for k sets.

$$\Gamma_1 \times \Gamma_2 \times \Gamma_3 \times \cdots \times \Gamma_k = \{(b_1, b_2, b_3, \ldots, b_k)/b_i \in \Gamma_i \text{ for all } i = 1, 2, 3, \ldots, k\}$$

Theorem 4.1 *If* $|H| = m$ *and* $|P| = k$ *then* $|H \times P| = |P \times H| = m \times k$.

Proof The proof is trivial.

Theorem 4.2 *If H, P, S are non-empty sets then* $H \subseteq P \Rightarrow H \times S \subseteq P \times S$

Proof Let (h, s) be any element in $H \times S$, then

$$\Rightarrow h \in H \subseteq P \text{ and } s \in S, \Rightarrow (h, s) \in P \times S, \text{ Hence } H \times S \subseteq P \times S.$$

Example 4.1 Find $H \times P, P \times S, P \times H$, where $H = \{111, 112, 113, 114\}$, $P = \{u, w\}$, $S = \{a, b, c, d\}$.

Solution We have,

$$\begin{aligned} H \times P &= \{(111, u), (111, w), (112, u), (112, w), (113, u), (113, w), (114, u), \\ &\quad (114, w)\}, \\ P \times S &= \{(u, a), (u, b), (u, c), (u, d), (w, a), (w, b), (w, c), (w, d)\}, \\ P \times H &= \{(u, 111), (u, 112), (u, 113), (u, 114), (w, 111), (w, 112), \\ &\quad (w, 113), (w, 114)\}, \end{aligned}$$

4.3 Relations: Experiential Learning

Inferences-II

Now, let's consider the various possibilities of the Experiment-I.

1. Suppose a student S_1 want to get admission at any college but the program must be Computer Science and every college is offering this program.
2. A student S_{10} wants to get admission in C_1 or C_5, or C_{10} colleges, but program could be anyone that is available in the college. Therefore, $A_{10} = \{(S_{10}, C_1), (S_{10}, C_5), (S_{10}, C_{10})\}$
3. A student S_2 and S_8 want to get admission in C_2 or C_3 colleges only. Therefore, the corresponding set is $A_{28} = \{(S_2, C_2), (S_2, C_3), (S_8, C_2), (S_8, C_3)\}$
4. A student S_8 wants to get admission in the IIT Bombay, but IIT Bombay is not in the set B. So, the corresponding set is $A_8 = \{\phi\}$

All these sets are the subsets of the set S given by 4.1, but their meanings are different. This leads to the concept of the Relation which is given in the next section.

Relation

Let H and P be two non-empty sets. A relation from H to P is any subset of the Cartesian Product $H \times P$. It is denoted by $R : H \to P$.

The above sets A_1, A_{10}, A_{28}, and A_8 are examples of the relation from A to B.

The elements of the relation set are defined as "An order pair (x, y) is in R means an element x in A is related to an element y in B." It is also denoted by xRy.

Important Results

(1) If (h, p) is in R then its meaning is h is related to p or hRp.
(2) The possible number of relations from H to P is $2^{m\times n}$, if the number of elements in H is m and P is n.
(3) R is a relation on the set H if $R : H \times H \rightarrow H$

4.4 Pictorial Representatives of Relations: Experiential Learning

Every relation can be signified pictorially by drawing its graph. Let H be any non-empty set and R be a relation on H. The relation R can be represented graphically by the following procedure.

(i) The elements of the set H are represented by small circles or point, i.e., (o) or these elements are called as vertices or nodes.
(ii) If $(h, m) \in R$ or hRm then vertices h and m are joined by a continuous arc with an arrow from h to m. These arcs are known as edges of the graph, i.e., if hRm then (Fig. 4.1)
(iii) If mRm then the vertex m is joined to itself by a loop around m. If mRm then (Fig. 4.2)

The graphical representation of the relation is called as digraph or directed graph of R.

Let us consider the following examples.

(1) $aRa \wedge aRb \wedge bRa \wedge bRb$ (Fig. 4.3)
(2) $aRb \wedge bRc \wedge aRc \wedge aRa$ (Fig. 4.4)

Example 4.3 Let $H = \{1, 2, 3, 4, 5, 6\}$ and hRm iff h divides m. Draw digraph of R.

Solution We have

$$R = \{(1, 1), (2, 2), (3, 3), (4, 4), (5, 5), (6, 6), (1, 2), (1, 3), (1, 5), (1, 6), (2, 4), (2, 6), (3, 6)\}$$

The diagram of R is as follows (Fig. 4.5):

Fig. 4.1 Digraph of *hRm*

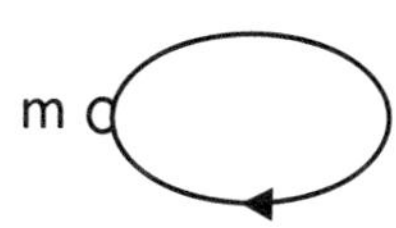

Fig. 4.2 Digraph of *mRm*

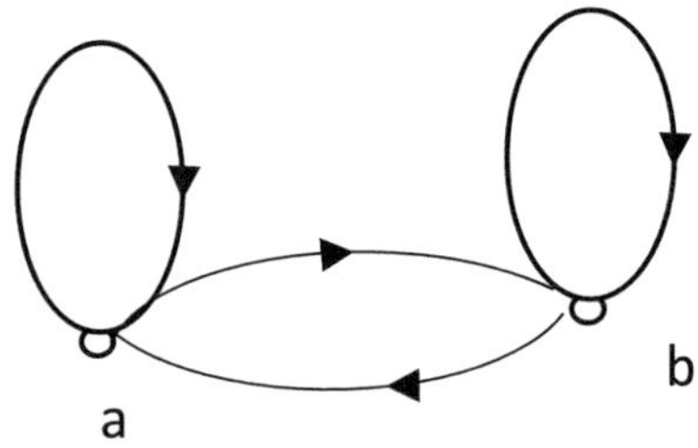

Fig. 4.3 Digraph of example (1)

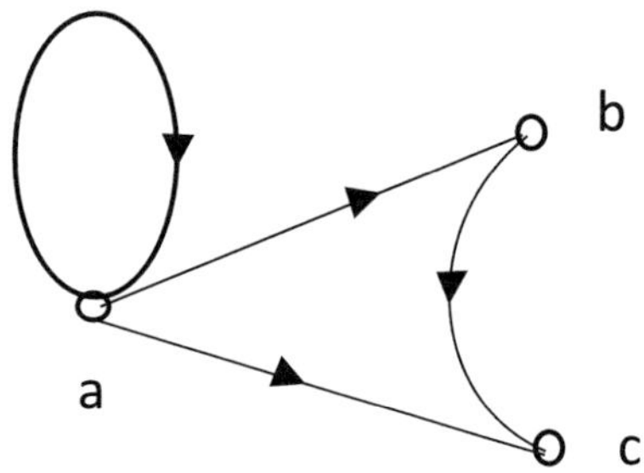

Fig. 4.4 Digraph of example (2)

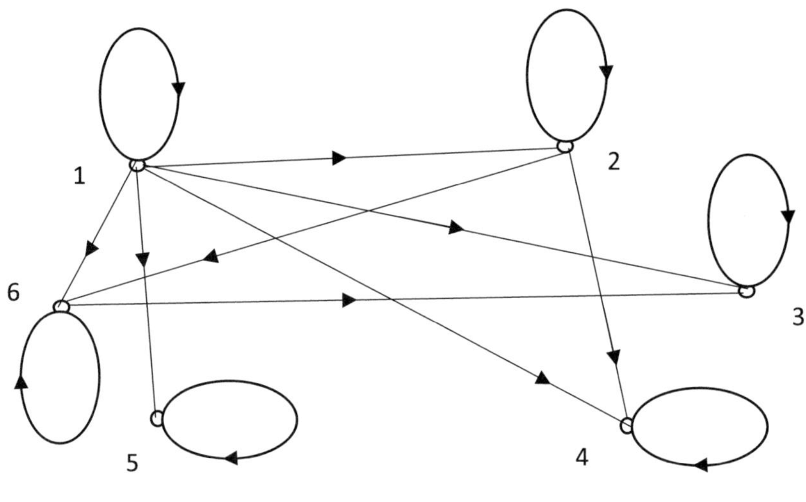

Fig. 4.5 Digraph of the relation R

4.4.1 Matrix Representation

Let $H = \{a_1, a_2, a_3, \ldots, a_n\}$, $P = \{b_1, b_2, b_3, \ldots, b_m\}$ and $R \subseteq H \times P$

Then the relation matrix of R is denoted by $M_R = \left[m_{ij}\right]_{n\times m}$ and defined by

$$m_{ij} = \begin{cases} 0 \text{ if } (a_i, a_j) \notin R \text{ i.e. } a_i R a_j \\ 1 \text{ if } (a_i, a_j) \in R \text{ i.e. } a_i R a_j \end{cases}$$

Example 4.4 Let $A = \{21, 22, 23, 24\}$ and $R = \{(h, m)/h < m\}$ then find M_R.

Solution $R = \{(21, 22)\ (21, 23)\ (21, 24)\ (22, 23)\ (22, 24)\ (23, 24)\}$

$$M_R = [M_{ij}]_{4\times 4} = \begin{matrix} & 21\ 22\ 23\ 24 \\ \begin{matrix} 21 \\ 22 \\ 23 \\ 24 \end{matrix} & \begin{bmatrix} 0 & 1 & 1 & 1 \\ 0 & 0 & 1 & 1 \\ 0 & 0 & 0 & 1 \\ 0 & 0 & 0 & 0 \end{bmatrix} \end{matrix}$$

Example 4.5 If $H = \{1, 2, 3, 4, 5, 6\}$ and hRp iff h divides p for $h, p \in H$. Find relation matrix.

Solution $R = \{(1, 2)\ (1, 3)\ (1, 4)\ (1, 5)\ (1, 6)\ (2, 4)\ (2, 6)\ (3, 6)\ (1, 1)\ (2, 2)\ (3, 3)$ $(4, 4)\ (5, 5)\ (6, 6)\}$

$$\text{Relation Matrix} = M_R = \begin{matrix} & 1\ 2\ 3\ 4\ 5\ 6 \\ \begin{matrix} 1 \\ 2 \\ 3 \\ 4 \\ 5 \\ 6 \end{matrix} & \begin{bmatrix} 1 & 1 & 1 & 1 & 1 & 1 \\ 0 & 1 & 0 & 1 & 0 & 1 \\ 0 & 0 & 1 & 0 & 0 & 1 \\ 0 & 0 & 0 & 1 & 0 & 0 \\ 0 & 0 & 0 & 0 & 1 & 0 \\ 0 & 0 & 0 & 0 & 0 & 1 \end{bmatrix}_{6\times 6} \end{matrix}$$

4.4.2 Relation Matrix Operations

We know that relation matrix is a Boolean matrix as all entries are either 0 or 1.

(I) Let $A = \left[a_{ij}\right]_{m\times n}, B = \left[b_{ij}\right]_{m\times n}$ be two relation matrices then $A + B = \left[a_{ij} + b_{ij}\right] = \left[c_{ij}\right]_{m\times n}$

$$\begin{aligned} \text{where } c_{ij} &= 1 \text{ if } a_{ij} = 1, \text{ or } b_{ij} = 1 \\ &= 0 \text{ if } a_{ij} = 0, \text{ and } b_{ij} = 0 \end{aligned}$$

(II) Let $A = \left[a_{ij}\right]_{m\times n,} B = \left[b_{jk}\right]_{n\times k}$ then $A.B = \left[a_{ij}; b_{jk}\right] = [d_{ik}]$

$$\begin{aligned} \text{where, } d_{ik} &= 1 \text{ if } a_{ij} = 1 \text{ and } b_{jk} = 1 \\ &= 0 \text{ if } a_{ij} = 0 \text{ and } b_{jk} = 0 \end{aligned}$$

e.g., If $A = \begin{bmatrix} 1 & 1 & 1 \\ 0 & 1 & 1 \\ 1 & 1 & 0 \end{bmatrix}$, $B = \begin{bmatrix} 1 & 1 & 1 \\ 0 & 1 & 1 \\ 0 & 0 & 0 \end{bmatrix}$ then $A + B = \begin{bmatrix} 1 & 1 & 1 \\ 0 & 1 & 1 \\ 1 & 1 & 0 \end{bmatrix}$ and $A.B = \begin{bmatrix} 1 & 1 & 1 \\ 0 & 1 & 1 \\ 1 & 1 & 1 \end{bmatrix}$.

4.5 Composite Relations

Let $R : H \to P$ and $S : P \to T$ be two relations. The Composition of R and S is denoted by $R \circ S$ or $RS : H \to T$ and defined as

$$R \circ S \text{ or } RS = \{(h, t)/(h, p) \in R \text{ and } (p, t) \in S \text{ for } h \in H, p \in P \text{ and } t \in T\}$$

Note: Let $R : H \to P$ and $S : P \to T$ be two relations. The Composition of R and S is denoted by $R \circ S$ or $RS : A \to C$ then $(R \circ S)^C = S^C \circ R^C$ [4, 5].

4.6 Types of Relations: Experiential Learning

A relation from a set H to itself has special significance in many applications. The $H \times H$ has n^2 elements, if $|H| = n$. We know that, the set having m number of elements has 2^m number of subsets. Therefore, there are 2^{n^2} relation are there on the set H. Let us consider the different types of the relations on the set H [5].

Brainstorming Experiment-III

Suppose, H is the set of 20 students of your class that includes 10 boys and 10 girls. Let's define the relation on the set H as follows.

$(h, p) \in R$ if and only if a student h is taking care of student p or h loves p or h is the best friend of p. Now, discuss with the students and find out all possible combinations of the cases.

The Cartesian product of H with H, if $H = \{s_i/s_i \text{ is a student and } i = 1, 2, 3, \ldots, 20\}$ is

$$H \times H = \left\{ \begin{array}{l} (s_1, s_1), (s_1, s_2), (s_1, s_3), \ldots, (s_1, s_{20}) \\ (s_2, s_1), (s_2, s_2), (s_2, s_3), \ldots, (s_2, s_{20}) \\ (s_3, s_1), (s_3, s_2), (s_3, s_3), \ldots, (s_3, s_{20}) \\ \vdots \\ (s_{20}, s_1), (s_{20}, s_2), (s_{20}, s_3), \ldots, (s_{20}, s_{20}) \end{array} \right\}.$$

The following are the list of possible cases.

(1) The possible cases for an individual student are given below.

(a) Every student loves himself. So, $H_1 = \{(s_i, s_i)/\text{for all } i = 1, 2, 3, \ldots, 20\}$
(b) No one loves oneself. Therefore, $H_2 = \{(s_i, s_i)/\text{for no } i = 1, 2, 3, \ldots, 20\}$
(c) Not all students love themselves. Suppose the first 10 students love themselves and others are not.
So, $H_3 = \{(s_i, s_i)/\text{for } i = 1, 2, 3, \ldots, 10\}$

(2) The probable cases for a pair of two students are as follows.

(a) A student s_i loves s_j if s_j loves s_i. Hence,

$$H_4 = \left\{(s_i, s_j) \in H_4/\text{whenever } (s_j, s_i) \in H_4 \text{ for some } i \text{ and } j\right\}$$

(b) A student s_i loves s_j but s_j does not loves s_i. Hence,

$$H_5 = \left\{(s_i, s_j) \in H_4/(s_j, s_i) \notin H_5 \text{ for some } i \text{ and } j\right\}$$

(3) The possible combinations or cases for three students are as follows.

(a) A student s_i loves s_j and s_j loves s_k then the student s_i loves s_k. Hence,

$$H_6 = \{(s_i, s_k) \in H_6/\text{whenever } (s_i, s_j) \in H_6 \text{ and } (s_j, s_k) \in H_6 \text{ for some } i, j \text{ and } k\}$$

(b) A student s_i loves s_j and s_j loves s_k but then the student s_i does not love s_k. Hence,

$$H_7 = \left\{(s_i, s_k) \notin H_6/(s_i, s_j) \in H_7 \text{ and } (s_j, s_k) \in H_7 \text{ for some } i, j \text{ and } k\right\}$$

Inferences-III

The following are the observed inferences of the above three possible cases.

1. Every individual student loves himself or doesn't love himself. {Cases (1): (a), (b) and (c)}.
2. If a student x loves y then y loves x or y doesn't love x. {Cases (2): (a) and (b)}.
3. If a student x loves y and y loves z then the student x loves z or the student x doesn't love student z.

These three inferences help us to define the types of the relations.

Mathematical Insights-III

Now, express the above inferences in terms of the mathematical theory or notations or symbols. The above three inferences suggest that there are mainly three types of relations.

1. Reflexive Relation
2. Symmetric Relations
3. Transitive Relations

Let's define one by one by using the above inferences and experiment.

4.6.1 *Reflexive Relation*

(i) A relation R on the set H is said to be **Reflexive Relation (RR)** if

$$(h, h) \in R \text{ or } hRh \text{ for all } h \text{ in } H$$

The example in the case (1) (a) is reflexive relation on the concerned set H.

(ii) It is **not reflexive relation** if $\exists h$ in H such that $(h, h) \notin R$.

The example in the case (1) (c) is not reflexive relation on the concerned set H.

(iii) The relation R on the set H is said to be **irreflexive** if $(e, e) \notin R$ for all e in H.

The example in the case (1) (b) is irreflexive relation on the set H.

Notes:

1. The Relation Matrix (M_R) of a reflexive relation is the Identity matrix.
2. Digraph of the reflexive relation has loop at every vertex or element of the set H.
3. The relation Matrix of irreflexive relation has all diagonal elements zero. Its digraph is free from loops.
4. The number of reflexive relations on H is 2^{n^2-n} if $|H| = n$

Example 4.6 Which of the following are Reflexive or not reflexive or irreflexive of the set $H = \{1, 2, 3, 4\}$.

i. $R_1 = \{(1, 1), (2, 2), (3, 3), (4, 4)\}$,
ii. $R_2 = \{(1, 1), (1, 2), (2, 2), (2, 1), (3, 3), (4, 4)\}$,
iii. $R_3 = \{(4, 4), (2, 3), (3, 4), (4, 1)\}$,
iv. $R_4 = \{(3, 5), (4, 5)\}$,
v. $R_5 = \{(3, 1), (4, 2)\}$.

Solution

i. There are only four elements in H and every element is related to itself in R_1. So, R_1 is a reflexive relation on H.
ii. The relation R_2 involves R_1 and some more elements. So, R_2 is reflexive relation.
iii. The relation R_3 is not reflexive relation as there is an element 2 in H but (2, 2) does not present in R_3.
iv. R_4 is not a relation on H as 5 is not in H.
v. R_5 is irreflexive relation as there is no element in R_5 that is related to itself.

4.6.2 *Symmetric Relation*

(i) A relation R on the set H is said to be **Symmetric relation** if whenever $(h, p) \in R$ then $(p, h) \in R$ for all h and p in H [4]
The example in the case (2) (a) is symmetric relation on the set H.
(ii) A Relation R is **not symmetric relation** if $\exists(h, p) \in R$ but $(p, h) \notin R$.
The Example in the case (2) (b) is not symmetric relation on the set H.
(iii) A Relation R is **asymmetric relation** if whenever $(h, p) \in R$ then $(p, h) \notin R$.

Notes:

1. The relation matrix of symmetric relation is a symmetric matrix.
2. A relation is symmetric iff $R = R^{-1}$.
3. The total number of symmetric relations on the set H having n elements is $2^{\frac{n(n+1)}{2}}$

4.6.3 *Transitive Relation*

(i) A relation R on the set H is said to be **Transitive Relation (TR)** if whenever $(h, p) \in R$ and $(p, m) \in R$ then $(h, m) \in R$ for all h, p and m in H.
(ii) A relation R is **not transitive relation** if $\exists$ at least one triplet in H such that $(h, p) \in R$ and $(p, m) \in R$ but $(h, m) \notin R$ [4].

Example 4.7 Which of the relations in Example 1 are transitive?

Solution

i. There are only four elements in H and every element is related to itself in R_1. So, R_1 is a transitive relation on H.
ii. R_2 is TR as every transitive triplet is present in the relation.
iii. The relation R_3 is not TR as (2, 3) and (3, 4) are in R_3 but (2, 4) does not present in R_3.
iv. R_4 is not a relation on H as 5 is not in H.
v. R_5 is transitive relation as it satisfies all conditions transitive relation.

4.6.4 *Special Types of Relations: Experiential Learning*

There are two types of the special relations depending upon the relation defined on one or two sets.

Special Types of Relations on One Set

Brainstorming Experiment-IV

Let $S = \{a, b, c, d, e\}$ be the set of five students of a class. Out of these five students, student b and c are honest with themselves or taking the desirable care of their health and family, but a, d, and e are not honest with themselves. Let's define the relations R on the set S as: x is related to y if x is honest with y. So,$(x, x) \in R$ iff x is honest with oneself. Therefore, $R = \{(b, b), (c, c)\}$ as b and c are honest with themselves. Hence, the relation R is the Identity relation of the set S.

Inferences-IV

The following are the notable observations of the above experiment.

- A set contains elements which are related to itself.
- A set contains elements which are related to itself as well as other elements.
- A set does not contain elements which are related to itself.

Mathematical Insights-IV

1. **Identity Relation**
   A relation R on a set H, is the Identity Relation if

$$R = \{(h, h)/\text{for}\, h \in H\}$$

Example 1: If $H = \{1, 2, 3, 4\}$ and $R_1 = \{(1, 1), (3, 3)\}$, $R_2 = \{(4, 4)\}$, $R_3 = \{(2, 2), (2, 3)\}$ then R_1 and R_2 are Identity relations. But R_3 is not the Identity relation as (2, 3) is present in R_3.

Brainstorming Experiment-V

Let $X = \{a, b, c\}$, and $P(X) = H$. Let S and M be two subsets of X. So, there are three possibilities.

Case 1: S and M are disjoint subsets.
Case 2: S and M have some common elements.
Case 3: All elements of S are in M as well as and all elements of M are in S.

Inferences-V

The following are the remarkable and important observations of the above experiment.

- $S \cap M = \phi$,
- $S \cap M \neq \phi$,
- $S \subseteq M$ and $M \subseteq S \Rightarrow S = M$. Let us understand this case through mathematical insights.

Mathematical Insights-V

2. **Antisymmetric Relation**
   A relation R on a set H is called antisymmetric relation if whenever (h, p) and (p, h) are in R then $h = p$.

A relation is not antisymmetric if $\exists h, p \in H$ such that $h \neq p$ but

$$(h, p) \in R \text{ and } (p, h) \in R$$

The relation matrix of the antisymmetric matrix never be symmetric matrix.

Examples:

Let $H = R$ (real-numbers set) and define the relation R on H as:

$$hRp \text{ iff } h \leq p.$$

Now, suppose hRp and pRh then $h \leq p$ and $p \leq h$. This implies $h = p$. Thus, R is antisymmetric relation. It is not symmetric relation.

(1) Let $H = \{1, 2, 3, 4\}$ then

- (a) $R_1 = \{(1, 1), (3, 3)\}$ is a symmetric as well as antisymmetric relations.
- (b) $R_2 = \{(1, 2), (2, 1)\}$ is a symmetric relation but not antisymmetric relation.
- (c) $R_3 = \{(1, 1), (1, 3), (2, 2)\}$ is antisymmetric relation but not symmetric relation.
- (d) $R_4 = \{(1, 1), (2, 3)(3, 2), (3, 4)\}$ is neither symmetric nor antisymmetric relations.

3. **Compatible Relation**

A relation R is said to be **Compatible relation**, if it is Reflexive and Symmetric relation.

The Relation Matrix of the compatible relation is a symmetric matrix with the diagonal elements 1.

Note: If R is a compatible relation on a set H and $|H| = n$ then $n \leq |R| \leq n^2$.

Example:

(1) Let $A = \{13, 23, 33\}$ then

- (a) $R_1 = \{(13, 13), (23, 23), (33, 33)\}$ is a compatible relation.
- (b) $R_2 = \{(13, 23), (23, 13)(33, 33)\}$ is a symmetric relation but not compatible relation.
- (c) $R_3 = \{(13, 13), (23, 23), (33, 33), (23, 33)\}$ is reflexive relation but not symmetric relation, hence it is not compatible relation.

(2) Let $H =$ set of all straight lines in the plane. A relation R on H is defined as

$$R = \{(h, p)/h \text{ is parallel to } p\}$$

The relation R is a compatible relation.

Special Types of Relations on Two Sets

Brainstorming Experiment-VI

Let H be the set of all people in the "ATHARVA" society. Let h and p be any two elements in H. h is related to p iff h is a friend of p.

Inferences-VI

The following are the remarkable and significant observations of the above experiment.

- h is a friend of p and p is also a friend of h.
- h is a friend of p but p is not a friend of h.
- There may be a set in which no one is the friend of each other

Let us understand this case through mathematical insights.

Mathematical Insights-VI

This would help us to define the Inverse relation and the Complement of a relation.

1. **Inverse Relation**
   If $R : H \to P$ is a relation then the Inverse relation $R^{-1} : P \to H$ and $R^{-1} = \{(p, h)/(h, p) \in R \text{ for all } h \in H \text{ and } p \in P\}$.
   That is if xRy then $yR^{-1}x$. It is also known as the converse relation and denoted by R^c.
2. **Complement of a Relation**
   Let $R : H \to P$ be a relation. The Complement relation of R is denoted by $\bar{R}$ or $R' : H \to P$ and defined as $\bar{R} = \{(h, p)/(h, p) \notin R \text{ for } h \in H \text{ and } p \in P\}$.

Notes:

4.7 Closure Properties: Experiential Learning

Brainstorming Experiment-VII

Let H be a set of all students of the S. Y. B.Tech. Class. Define a relation R as hRp iff h is honest with p, for all h and p in H. Now, ask few questions to students.

1. How many of you are honest to yourself?
2. Whether your friend is honest with you or not even though you are honest with your friend?
3. If h is honest with p and p is honest with m then whether h is honest with m or not?

I think all students would be agreeing with me that for the smooth harmony of the class or institute, the answers of all above questions must be "YES." But, if a particular student is not agreeing with this, then we have to discuss with him and try to change his opinion. This leads to the adding the missing property to the set.

Inferences-VII

The following are the special observations of the above experiment.

- If few students are not honest with themselves then do counseling and make it happen that all students will be honest with themselves.
- If x is honest with y but y is not honest with x in particular cases, then counsel y, so that y would be honest with x.
- If x is honest with y and y is honest with z then x must be honest with z. All x, y, and z must honest with each other.

Let us understand this case through mathematical insights.

Mathematical Insights-VII

The above inferences lead to the concepts of the reflexive closure, symmetric closure, and transitive closure. Let us define each of these concepts individually.

Closure of a Relation

Depending upon the property under consideration, there are three types of the closures.

4.7.1 *Reflexive Closure*

The reflexive closure (RC) of the relation S on the set H is

$$RC = S \cup \Delta$$

where $\Delta = \{(d, d)/\forall d \in H\}$

If $H = \{aa, ba, ca, da\}$ and R is any relation on H then $\Delta = \{(aa, aa)\,(ba, ba)\,(ca, ca)\,(da, da)\}$ and $R \cup \Delta$ is reflexive closure of R [5].

Example 4.8 Let $H = \{11, 12, 13\}$ and R_1, R_2, and R_3 are relations on set H. Find the reflexive closures of R_1, R_2, and R_3, where $R_1 = \{(11, 11)\,(12, 11)\}$, $R_2 = \{(11, 11)\,(12, 12)\,(13, 13)\}$, $R_3 = \{(13, 11)\,(11, 13)\,(12, 13)\}$.

Solution We have $H = \{11, 12, 13\}$ and $\Delta = \{(11, 11)\,(12, 12)\,(13, 13)\}$.

Therefore,

(i) The reflexive closure (RC) of R_1 is $R = R_1 \cup \Delta = \{(11, 11)\,(12, 12)\,(13, 13)\,(12, 11)\}$,
(ii) The RC of R_2 is $R = R_2 \cup \Delta = R_2$,
(iii) The RC of R_3 is $R = R_3 \cup \Delta = \{(11, 11)\,(12, 12)\,(13, 13)\,(13, 11)\,(11, 13)\,(12, 13)\}$.

4.7.2 Symmetric Closure

A Symmetric Closure (SC) of a relation R is

$$\text{SC} = R \cup R^{-1}$$

Example 4.9 Find the SC of the following, where $H = \{1, 2, 3\}$.

$$R_1 = \{(1, 1)\ (2, 1)\},$$
$$R_2 = \{(1, 2)\ (2, 1)\ (3, 2)\ (2, 2)\},$$
$$R_3 = \{(1, 1)\ (2, 2)\ (3, 3)\}.$$

Solution Given that, $H = \{1, 2, 3\}$.

(i) $\text{SC} = R_1 \cup R_1^{-1} = \{(1, 1)\ (1, 2)\ (2, 1)\}$,
(ii) $\text{SC} = R_2 \cup R_2^{-1} = \{(1, 2)\ (2, 1)\ (3, 2)\ (2, 3)\ (2, 2)\}$,
(iii) $\text{SC} = R_3$.

4.7.3 Transitive Closure (TC)

A relation R^* is transitive closure of the relation R if R^* is the smallest transitive relation such that $R \subseteq R^*$ [6, 7].

Theorem 4.4 *Let S be a relation on H, $S^* = S \cup S^2 \cup \cdots \cup S^n$.*

4.7.4 Warshall's Algorithm to Find Transitive Closure

Consider the following steps to find R^* of the relation R on a set H [8, 9].

Step 1: We have $|H| = n$.
$\therefore W_{0,}\ W_{1,}\ W_2, \ldots, W_n$ called as Warshall's sets.

$$W_0 = M_R$$

Step 2: To find R^*
Compute W_k from W_{k-1}.

(i) If $w_{st} = 1$ in W_{k-1} then in W_k, $w_{st} = 1$.
(ii) In column C_q, $w_{pq} = 1$ for $p = p_1, p_2, \ldots, p_a$ and in row Rq, $w_{qv} = 1$ for $v = v_1, v_2, \ldots, v_s$
then put 1's at the places
$(p_i, q_i) = 1$ for $i = 1, \ldots, a$ and $j = 1, \ldots, s$

Step 3: Apply the procedure n times to find W_n. Then, write R^* from W_n.

Example 4.10 Find the R^* by Warshall's algorithm.

$$H = \{1, 2, 3, 4, 5, 6\} \text{ and } R = \{(x, y)/|x - y| = 2\}$$

Solution Consider the following steps.

Step 1: we have $|H| = 6$

$$R = \{(1, 3)\ (3, 1)\ (2, 4)\ (4, 2)\ (4, 6)\ (6, 4)\ (3, 5)\ (5, 3)\}$$

Find W_0, W_1, W_2, W_4, W_5, and W_6.
The first set W_0 is same as M_R which is shown below

$$W_0 = M_R = \begin{array}{c} \\ 1 \\ 2 \\ 3 \\ 4 \\ 5 \\ 6 \end{array} \begin{array}{c} \begin{array}{cccccc} 1 & 2 & 3 & 4 & 5 & 6 \end{array} \\ \begin{bmatrix} 0 & 0 & 1 & 0 & 0 & 0 \\ 0 & 0 & 0 & 1 & 0 & 0 \\ 1 & 0 & 0 & 0 & 1 & 0 \\ 0 & 1 & 0 & 0 & 0 & 1 \\ 0 & 0 & 1 & 0 & 0 & 0 \\ 0 & 0 & 0 & 1 & 0 & 0 \end{bmatrix} \end{array}$$

Step 2: To find W_1:
To find W_1 from W_0, we consider the first column and first row.
In a column C_1, 1 is present at R_3.
In a row R_1, 1 is present at C_3.
Thus, add new entry in W_1 at (R_3, C_3) which is given below

$$W_1 = \begin{array}{c} \\ 1 \\ 2 \\ 3 \\ 4 \\ 5 \\ 6 \end{array} \begin{array}{c} \begin{array}{cccccc} 1 & 2 & 3 & 4 & 5 & 6 \end{array} \\ \begin{bmatrix} 0 & 0 & 1 & 0 & 0 & 0 \\ 0 & 0 & 0 & 1 & 0 & 0 \\ 1 & 0 & 1 & 0 & 1 & 0 \\ 0 & 1 & 0 & 0 & 0 & 1 \\ 0 & 0 & 1 & 0 & 0 & 0 \\ 0 & 0 & 0 & 1 & 0 & 0 \end{bmatrix} \end{array}$$

Step 3: To find W_2:
To find W_2 from W_1, we consider the second column C_2 and second row R_2.
In a column C_2, 1 is present at R_4.
In a row R_2, 1 is present at C_4.

Thus, add new entry in W_2 at (R_4, C_4) which is given below

$$W_2 = \begin{array}{c} \\ 1 \\ 2 \\ 3 \\ 4 \\ 5 \\ 6 \end{array} \begin{array}{c} \begin{array}{cccccc} 1 & 2 & 3 & 4 & 5 & 6 \end{array} \\ \begin{bmatrix} 0 & 0 & 1 & 0 & 0 & 0 \\ 0 & 0 & 0 & 1 & 0 & 0 \\ 1 & 0 & 1 & 0 & 1 & 0 \\ 0 & 1 & 0 & 1 & 0 & 1 \\ 0 & 0 & 1 & 0 & 0 & 0 \\ 0 & 0 & 0 & 1 & 0 & 0 \end{bmatrix} \end{array}$$

Step 4: To find W_3:

To find W_3 from W_2, we consider the third column and third row.

In a column C_3, 1 is present at R_1, R_3, R_5

In a row R_3, 1 is present at C_1, C_3, C_5

Thus add new entry in W_3 at (R_1, C_1) (R_1, C_3) (R_1, C_5) (R_3, C_1) (R_3, C_3) (R_3, C_5) (R_5, C_1) (R_5, C_3) (R_5, C_5) which is given below

$$W_3 = \begin{array}{c} \\ 1 \\ 2 \\ 3 \\ 4 \\ 5 \\ 6 \end{array} \begin{array}{c} \begin{array}{cccccc} 1 & 2 & 3 & 4 & 5 & 6 \end{array} \\ \begin{bmatrix} 1 & 0 & 1 & 0 & 1 & 0 \\ 0 & 0 & 0 & 1 & 0 & 0 \\ 1 & 0 & 1 & 0 & 1 & 0 \\ 0 & 1 & 0 & 1 & 0 & 1 \\ 1 & 0 & 1 & 0 & 1 & 0 \\ 0 & 0 & 0 & 1 & 0 & 0 \end{bmatrix} \end{array}$$

Step 5: To find W_4:

To find W_4 from W_3, we consider the fourth column and fourth row.

In a column C_4, 1 is present at R_2, R_4, R_6

In a row R_4, 1 is present at C_2, C_4, C_6

Thus, add new entry in W_4 at (R_2, C_2) (R_2, C_4) (R_2, C_6) (R_4, C_2) (R_4, C_4) (R_4, C_6) (R_6, C_2) (R_6, C_4) (R_6, C_6) which is given below

$$W_4 = \begin{array}{c} \\ 1 \\ 2 \\ 3 \\ 4 \\ 5 \\ 6 \end{array} \begin{array}{c} \begin{array}{cccccc} 1 & 2 & 3 & 4 & 5 & 6 \end{array} \\ \begin{bmatrix} 1 & 0 & 1 & 0 & 1 & 0 \\ 0 & 1 & 0 & 1 & 0 & 1 \\ 1 & 0 & 1 & 0 & 1 & 0 \\ 0 & 1 & 0 & 1 & 0 & 1 \\ 1 & 0 & 1 & 0 & 1 & 0 \\ 0 & 1 & 0 & 1 & 0 & 1 \end{bmatrix} \end{array}$$

Step 6: To find W_5:

To find W_5 from W_4, we consider the fifth column and fifth row.

In a column C_5, 1 is present at R_1, R_2, R_3

In a row R_5, 1 is present at C_3

Thus, add new entry in W_5 at (R_1, C_3) (R_2, C_3) (R_3, C_3) which is given below

$$W_5 = \begin{array}{c} \\ 1 \\ 2 \\ 3 \\ 4 \\ 5 \\ 6 \end{array} \begin{array}{c} \begin{array}{cccccc} 1 & 2 & 3 & 4 & 5 & 6 \end{array} \\ \begin{bmatrix} 1 & 0 & 1 & 0 & 1 & 0 \\ 0 & 1 & 0 & 1 & 0 & 1 \\ 1 & 0 & 1 & 0 & 1 & 0 \\ 0 & 1 & 0 & 1 & 0 & 1 \\ 1 & 0 & 1 & 0 & 1 & 0 \\ 0 & 1 & 0 & 1 & 0 & 1 \end{bmatrix} \end{array}$$

Step 7: To find W_6:

To find W_6 from W_5, we consider the sixth column and sixth row.

In a column C_6, 1 is present at R_2, R_4, R_6

In a row R_6, 1 is present at C_2, C_4, C_6

Thus, add new entry in W_6 at (R_2, C_2) (R_2, C_4) (R_2, C_6) (R_4, C_2) (R_4, C_4) (R_4, C_6) (R_6, C_2) (R_6, C_4) (R_6, C_6) which is given below

$$W_6 = W_5 = W_4$$

Hence, W_6 is the relation matrix of R^*

$$R^* = \begin{Bmatrix} (1,1)(1,3)(1,5)(2,2)(2,4)(2,6)(3,1)(3,3)(3,5) \\ (4,2)(4,4)(4,6)(5,1)(5,3)(5,5)(6,2)(6,4)(6,6) \end{Bmatrix}$$

Example 4.11 Find R^* where

$A = \{51, 52, 53, 54\}$ defined by

$$R = \{(51,52)(51,53)(51,54)(52,51)(52,53)(53,54)(53,52)(54,52)(54,53)\}$$

Solution By applying all steps of the Warshall's Algorithm, the transitive closure of the given relation is given below.

$$R^* = \begin{Bmatrix} (51,51)(51,52)(51,53)(51,54)(52,51)(52,52)(52,53)(52,54) \\ (53,51)(53,52)(53,53)(53,54)(54,51)(54,52)(54,53)(54,54) \end{Bmatrix}.$$

Example 4.12 Find R^* where

$$M_R = \begin{bmatrix} 1\ 0\ 1 \\ 0\ 1\ 0 \\ 1\ 1\ 0 \end{bmatrix} \text{ and } H = \{10, 20, 30\}.$$

Solution $R^* = \{(10, 10)(10, 20)(10, 30)(20, 20)(30, 10)(30, 20)(30, 30)\}$.

4.7.5 *Bhapkar's Matrix Method (BMM) for Finding the Transitive Closure*

Let us consider the following algorithm for finding the transitive closure of a non-transitive relation R on the set H.

Step 1: Let H be any set having n elements. Write the matrix representation $M(R)$ of the relation R. The size of this matrix is $n \times n$.
Step 2: Select the ith row for $i = 1, 2, 3, \ldots, n$. Find the first position having 1 in $M(R)$ say (i, j)th position.
Step 3: Put all 1's of the jth row at the corresponding places of the ith row having zero elements. Keep all 1's of the ith row as it is.
Step 4: Apply step 2 and step 3 for all elements of the first row again. If all such elements are added at the first row.
Step 5: Apply steps 2 to 4 for all i and j.
Step 6: We will get the matrix of the required transitive relation. Write the transitive relation R^* form the matrix.

Example 4.13 Let $H = \{1, 2, 3, 4\}$ defined by $R = \{(1, 2), (1, 3), (1, 4), (2, 1), (2, 3), (3, 4), (3, 2), (4, 2), (4, 3)\}$. Find the Transitive closure of R by BMM Method.

Solution Consider the following steps.

1. The matrix of given relation is given as follows.

$$M(R) = \begin{bmatrix} 0\ 1\ 1\ 1 \\ 1\ 0\ 1\ 0 \\ 0\ 1\ 0\ 1 \\ 0\ 1\ 1\ 0 \end{bmatrix}$$

2. In the first row, the first 1 is at position (1, 2), so write all 1's of the second row at corresponding position of the first row of $M(R)$. Similarly, write all 1's of the third and fourth rows at the first row and denote the revised matrix by H_1

$$H_1 = \begin{bmatrix} 1 & 1 & 1 & 1 \\ 1 & 0 & 1 & 0 \\ 0 & 1 & 0 & 1 \\ 0 & 1 & 1 & 0 \end{bmatrix}$$

3. In the second row, the first 1 is at the place (2, 1), so write all 1's of the first row at the corresponding places of the second row. Now, add all 1's of the third row also at the second row. The revised matrix is denoted by H_2.

$$H_2 = \begin{bmatrix} 1 & 1 & 1 & 1 \\ 1 & 1 & 1 & 1 \\ 0 & 1 & 0 & 1 \\ 0 & 1 & 1 & 0 \end{bmatrix}$$

Applying the similar process for all remaining rows and 1's, we get the following matrices.

$$H_3 = \begin{bmatrix} 1 & 1 & 1 & 1 \\ 1 & 1 & 1 & 1 \\ 1 & 1 & 1 & 1 \\ 0 & 1 & 1 & 0 \end{bmatrix}, H_4 = \begin{bmatrix} 1 & 1 & 1 & 1 \\ 1 & 1 & 1 & 1 \\ 1 & 1 & 1 & 1 \\ 1 & 1 & 1 & 1 \end{bmatrix} = M(R^*)$$

H_4 is the matrix of the transitive closure the given relation. Thus, the required TC of R is as follows

$$R^* = \left\{ \begin{array}{l} (1,1)(1,2)(1,3)(1,4)(2,1)(2,2)(2,3)(2,4) \\ (3,1)(3,2)(3,3)(3,4)(4,1)(4,2)(4,3)(4,4) \end{array} \right\}$$

Note: As compared with all existing method for finding the TC of the relation, the Bhapkar's Matrix Method is simple and easy for calculations as well understandings.

4.8 Equivalence Relation: Experiential Learning

Brainstorming Experiment-VIII

In India, there are many political parties working with their philosophy. Let H be the set of all voters in a particular constituency and suppose four candidates are contesting the election. Let C_1, C_2, C_3, C_4 and $V_1, V_2, V_3, \ldots, V_n$ be the candidates and voters, respectively. Every candidate is also a voter. So there are $n + 4$ voters. Now, define the relation as x is related to y iff x and y cast their vote to the same candidate or party. After the election what would be the possible results.

Inferences-VI

The following are the special observations of the above experiment.

- Every candidate has casted his vote to himself.
- Every candidate has received the good number of votes.
- Every voter must cast his vote to only one candidate.
- Voters x and x cast their vote to the same candidate. Therefore, this relation is reflexive.
- If voters x and y cast their vote to the candidate C_1 then y and x also cast their votes to C_1 only. So, this relation is Symmetric relation.
- If voters x and y cast their votes to the same candidate and y and z also cast their votes to the same candidate. So, x, y, and z cast their votes to the same candidate. This means, if x is related to y and y is related to z then x is related to z. Hence, this relation is transitive relation.
- Now, let us divide all voters among parties as per their votes. All Voters of the candidate C_1 belong to the group of C_1 and similarly for other candidates. Therefore, there are only four disjoint groups. We can call them the party supporters, or in Mathematics language, it is an equivalence class of the party or candidate.
- The candidate having maximum number of elements in his equivalence class will be the winner of the election. Thus, political parties should increase their equivalence classes to win all types of elections.

Let us understand this case through mathematical insights.

Mathematical Insights-VIII

The above experiment would help us to define an equivalence relation (ER) and equivalence classes.

Equivalence Relation

A relation R on the set H is an equivalence relation (ER) if it is reflexive, symmetric, and transitive relation.

In an equivalence relation, $hRp = h \sim p$.

The experiment-VI give an equivalence relation for the defined relation.

Example 4.18 Let H be all straight lines in a plane. Let $R = \{hRp/h$ is parallel to $p\}$, $S = \{hSp/h$ is perpendicular to $p\}$. Check whether R and S are equivalent relations or not.

Solution The relation R is ER, but S is not.

Equivalence Class

Let R be an ER on a set H. The equivalence class of an element x in H is denoted by $[x]_R$ and defined as $[x]_R = \{y \in H/xRy$ for all $y \in H\}$.

In experiment-VI, the set all voters form an equivalence class with respect to the equivalence relation defined in the experiment.

Let's consider the live experiment of your institution/college/department. Let H bet the set of all students studying in your institution. Define relation R as a student x is related to student y iff x and y are studying in the same class or division. We have, student x and x are studying in the same class. If x and y are studying in the same class then y and x are also studying in the same class. Moreover, if students x and y are studying in the same class and y and z are studying in the same class then x and z are also studying in the same class. Therefore, the relation R is reflexive, symmetric and transitive relations. Hence, the relation R is an equivalence relation.

Let x be any student of the institution. The equivalence class of x is the set of students studying in his class. If x is studying in the class S. Y. B.Tech. Computer Engineering, division-B then the set all students studying in the same division is the equivalence class of the student x. Every students belongs to his own equivalence class. Every division of the institution is the equivalence class. So, the number of equivalence classes is equal to the number of divisions in the institution. Every division must have at least one student. This means, every equivalence class must have at least one element.

Properties of an Equivalence Class

1. Every element is a member of its own equivalence class. That is, $x \in [x]_R$
2. Every element is present in only one equivalence class.
3. Every equivalence class has at least one element.
4. If there are n elements in the set H and R is an equivalence relation on H, then there are at most n equivalence classes with respect to the relation R.

Partitions of a Set

We shall now discuss the concept of partitions of a set which is similar to equivalence class of set.

Definition: Let H be any non-empty set. A set $P = \{H_1, H_2, H_3, \ldots, H_n\}$ of non-empty subsets of H is called a partition of set H if

(i) $H_1 \cup H_2 \cup H_3 \cup \cdots \cup H_n = H = \bigcup_{i=n}^{n} H_i$
(ii) $H_i \cap H_j = \phi$ for $i \neq j$, i.e., All sets H_i are mutually disjoints.
The partition of a set H is denoted by π.

An element of a partition set is called a block. The rank of a partition is called as the number of blocks of that partitions. It is denoted by $r(\pi)$.

e.g.,

(1) Let $H = Z =$ set of all integers. Then, $H_1 =$ set of even integers, $H_2 =$ set of odd integers

$$H_3 = \{1, 2, 3, 4 \ldots\}\ H_4 = \{0\}\ H_5 = \{-1, -2, -3, -4, \ldots\}$$

$P_1 = \{H_1, H_2\}$ and $P_2 = \{H_3, H_4, H_5\}$ are different partitions of set H.

(2) If $A = \{1, 2, 3, 4, 5, 6, 7, 8, 9, 10\}$ and it's subsets

$$A_1 = \{1, 4, 9\} \quad A_2 = \{2, 6, 8, 10\} \quad A_3 = \{3, 5, 7\}$$

Product and Sum of Partitions

(I) Let π_1 and π_2 be two partitions of a non-empty set H. The product of π_1 and π_2 is denoted by Let $\pi_1.\pi_2 = \pi$ is a partition π of H such that

(1) π refines both π_1 and π_2
(2) If π^1 refines both π_1 and π_2 then π^1 refines π.

(II) Let π_1 and π_2 be the partitions of a non-empty set H. The sum of π_1 and π_2 is denoted by $\pi = \pi_1 + \pi_2$ is a partition of A such that

(i) Both π_1 and π_2 refine π
(ii) If π_1 and π_2 refine π^1 then π refines π^1.

4.9 Partial Ordering Relations: Experiential Learning

Brainstorming Experiment-IX

Let H be the set of five types of the businessmen having different gross annual income as depicted in the following table.

Businessmen (h_i)	h_1	h_2	h_3	h_4	h_5
Gross annual income in crore (G_i)	50	100	0.1	0.3	10

Let h_i and h_j be any two elements in H and h_i is related to h_j iff the gross income of h_i is less than or equal to the gross income of h_j, i.e., $h_i\ Rh_j$ iff $G_i \leq G_j$.

Inferences-IX

The following are the remarkable and significant observations of the above experiment.

i. $G_i \leq G_i$ for all i.
ii. If $G_i \leq G_j$
iii. If $G_i \leq G_j$ and $G_j \leq G_i$ then $G_i = G_j$
iv. If $G_i \leq G_j$ and $G_j \leq G_k$ then $G_i \leq G_j \leq G_k$

Let us understand this case through mathematical insights.

Mathematical Insights-IX

The following are the mathematical insights for the above inferences.

i. h_iRh_i, i.e., R is reflexive relation
ii. If h_iRh_j then $h_j\not{R}h_i$ (h_j is not related to h_i).

iii. If $h_i R h_j$ and $h_j R h_i$ then $h_i = h_j$. Therefore, R is antisymmetric relation.
iv. If $h_i R h_j$ and $h_j R h_k$ then $h_i R h_k$. Hence, R is a transitive relation.

Thus, R is reflexive, antisymmetric, and transitive relation. These types of properties lead to the concept of the Partially Ordered Set (POSET).

Partially Ordered Set

A relation R on a set H called a **Partially Ordered Relation** iff R is reflexive, antisymmetric, and transitive relation.

The set H together with partially ordered relation is called a Partially Ordered Set or **POSET**.

It is denoted by (H, R) or $(H, \leq)$, where $\leq$ is a partially ordered relation [10–12].

Examples:

1) $(\Re, \leq)$ $(N, \leq)$ are POSETS.
2) $(P(S), \subseteq)$ is a POSET.

I) **Comparable elements**: Let $(H, \leq)$ be a POSET. Two elements a, b in H are comparable elements if $a \leq b$ or $b \leq a$. Two elements a and b of a set H are non-comparable if neither $a \leq b$ nor $b \leq a$.

II) **Totally Ordered Set**: Let H be any non-empty set. The set H is called Linearly Ordered Set or Totally Ordered Set if $a \leq b$ or $b \leq a$, for all $a, b \in H$.

4.10 Case Studies and Projects

All readers are instructed to design your own case study/projects of the following situations.

1. Relations in your family/society
2. Relations of banks with their costumers
3. Relation of social workers and society
4. Relation of political leaders and the community
5. Relation of teachers and students
6. Relation of tree and people
7. Relation of students and library
8. Relation of friendship among students' groups
9. Relation among all teachers and subjects
10. Relation among company workers and their incentives.

Exercise

1. Define the following types and design your own example of each definition.

   i. Equivalence relation
   ii. Transitive closure
   iii. POSET

2. Construct two examples of the following relations

   i. Reflexive but not symmetric
   ii. Symmetric but not transitive
   iii. Transitive but not reflexive
   iv. Reflexive but not transitive and symmetric
   v. Antisymmetric but not symmetric
   vi. Symmetric but not antisymmetric
   vii. Equivalence but not antisymmetric

3. Find the Transitive closure of the following relations by Bhapkar's Matrix Method and Warshall's Algorithm on the set $H = \{p, q, r, s\}$

   i. $R_1 = \{(p, q), (q, r), (r, s)\}$,
   ii. $R_1 = \{(p, p), (q, s), (r, p), (s, r)\}$,
   iii. $R_1 = \{(p, q), (q, r), (r, s), (p, p), (q, q), (r, r), (s, s)\}$.

4. Find the reflexive and symmetric closures of the relations given in Example (3).
5. Let H be the set of all lines in the plane. Find the equivalence class of the line x in H if exists for the following relations.

   i. xRy iff x is $\|$ to y. (parallel)
   ii. xRy iff x is perpendicular to y.
   iii. xRy iff x intersects y.

6. If $H = \{1, 2, 3, 4, 5, 6, 7, 8, 9\}$ and $R = \{(h, p)/h - p$ is divisible by $3\}$. Is R an equivalence relation? If yes, find the equivalence class of all elements in H and draw the diagraph of R.
7. Draw the Hasse Diagram of a poset $(P(H), \subseteq)$, where $H = \{p, q, r, s\}$.
8. Explain any two applications of the Hasse Diagram.
9. Explain any five applications of equivalence relations.
10. Draw the mind map of this chapter.

References

1. Aigner M (2023) Discrete mathematics. American Mathematical Society
2. Erciyes K (2021) Discrete mathematics and graph theory. Springer, Cham. https://doi.org/10.1007/978-3-030-61115-6
3. Rosen K, Krithivasan K (2021) Discrete mathematics and its applications. Mc Graw Hill India. ISBN-978930727353
4. Jiri M, Jaroslav N (2011) Invitation to discrete mathematics, 2nd edn. Oxford University Press. ISBN-9780198570424
5. Tremblay JP, Manohar R (2023) Discrete mathematical structures with applications to computer science. McGraw Hill Education. ISBN-9780074631133
6. Martin DP (2017) Dynamic shortest path and transitive closure algorithms: a survey. arXiv preprint arXiv:1709.00553

7. Ainia A, Salehipour A (2012) Speeding up the Floyd-Warshall algorithm for the cycled shortest path problem. Appl Math Lett 25(1):1–5. https://doi.org/10.1016/j.aml.2011.06.008
8. Wang Q, Zhang D (2006) A simple and direct algorithm for computing transitive closure of fuzzy matrix. J Xi'an Univ Technol. en.cnki.com.cn/Article_en/CJFDTOTAL-XALD200603011.htm
9. Gupta AS, Pathak A (2020) Quantum Floyd-Warshall algorithm. arXiv:quant-ph/0502144
10. Baker KA, Fishburn PC, Roberts FS (1971) Partial orders of dimension 2. Networks 2(1):11–28. https://doi.org/10.1002/net.3230020103
11. Birkhoff G (1948) Lattice theory, rev edn. American Mathematical Society
12. Singh MK (2024) Chapter 1: fundamental of crisp set theory. Springer Science and Business Media LLC

Chapter 5
Functions

Mathematics is the language with which God has written the universe—Galileo Galilei

Abstract This chapter deals with a special type of relations on elements of two sets. A real relationship of children and mother is used as the brainstorming experiment to explain the function and types of functions such as one-to-one functions, onto functions, into functions, and bijective functions. Moreover, the particular examples support for the better understanding of applied aspects of the functions. It also focuses on the concepts of floor and ceiling function with examples. By considering the experiment of writing English words, this chapter explains the definitions of infinite sets, countable sets and related results with illustrations. The pigeonhole principle is enlightened by using the real examples of library systems in various institutions. Moreover, special examples are designed to understand the applications of this concept. Furthermore, discrete numeric functions are described with illustrations. This chapter concludes with the case studies, projects, and exercises to fill up the gap between theoretical concepts and their multifaceted applications.

Keywords Function · Bijective function · Floor and ceiling functions · Pigeonhole principle · Discrete numeric functions

5.1 Introduction

Chapter 4 explained about all types of relationships among elements of any two sets, their types, examples, and applications. The concept of function is associated with a particular types of relations among elements of two sets. However, the dictionary meaning of the function is "the purpose or special task or responsibility of things." The mathematical meaning of the function is also same as that of the dictionary. But, as Mathematics itself is a precise language, we express the function in mathematical notations with the same meaning [1, 2].

© The Author(s), under exclusive license to Springer Nature Singapore Pte Ltd. 2026

H. R. Bhapkar and P. N. Mahalle, *Discrete Mathematics: An Experiential Learning Perspective*, Studies in Autonomic, Data-driven and Industrial Computing,
https://doi.org/10.1007/978-981-95-7210-6_5

The development of the concept of function is similar with other mathematical concepts or theories. The growth of the function is as follows [3].

Year	Name of the mathematician	Descriptions
1673	Leibnitz	He used the term function first time to explain a quantity related to a curve such as slope or the coordinates. He also defined the terms Constant, parameter and variable
1694	Johann Bernoulli	He used the term function to describe a quantity formed from constant and indeterminate quantities
1748	Leonhard Euler	He defined function as an expression or formula that includes variables and constants
1837	Peter Gustav Lejeune Dirichlet	He defined function by using the relations

In this chapter, the function is defined by using relations only. Almost all terms of the pure Mathematics depend upon the concept of the function. So, it has numerous applications in every field of science, technology, and engineering [4, 5].

Let us use our "BEIMI" approach to understand the definition of function, types of functions, countability of sets, pigeonhole principle, and some discrete numeric functions.

5.2 Functions and Partial Functions: Experiential Learning

Let's consider the following brainstorming Experiment-I for understanding the concept of the Functions and types of Functions.

(A) **Brainstorming Experiment-I**

Let C be the set of p children in the Atharva Society. Let M be the set of q mothers of the children from the same society, such that the mother of each child is in the set M. We know that every child has only one biological mother. In this chapter, the mother refers to the biological mother of the child. Now, define a relation R from the set C to the set M as c is related to m, i.e., cRm or $(c, m) \in R$ iff c is a child of mother m, i.e., m is the mother of child c. According to the definition of the relation R, every child c_i is related to unique m_j. Therefore, this relation R is the special type of relation. The following set R denotes the special relation.

$$S = \{(c_1, m_{j1}), (c_2, m_{j2}), \dots (c_p, m_{jp})\} \tag{5.1}$$

where $m_{j1}, m_{j2}, \dots, m_{jp} \in M$

Let's consider the following inferences or observations.

Fig. 5.1

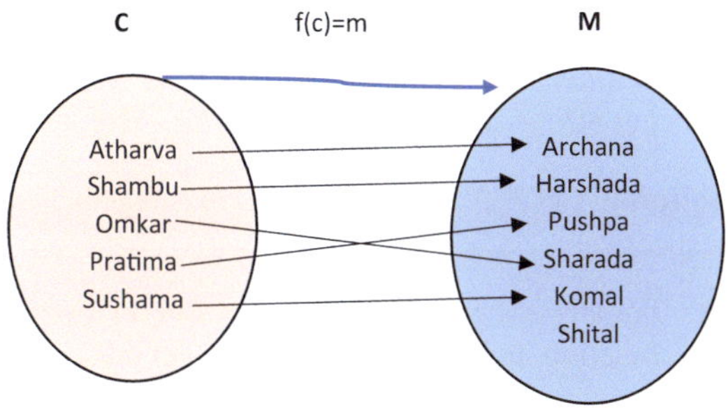

(B) **Inferences-I**

In Experiment-I, we have observed the following probable special inferences.

1. Every child c_i is associated with the unique mother m_j. Therefore, this is the special type of relation which defines the new concept "*FUNCTION.*"
2. In the first set, there may be some children whose mothers are not in the set M. Therefore, remove such children and define the function for a subset of C. It is known as the "*PARTIAL FUNCTION*."

Besides the above observations, there are numerous remarkable observations which will be explained according to the necessity of mathematical concept.

(C) **Mathematical Insights-I**

Functions

Let C and M be two non-empty sets. A relation from C to M is called as a Function if every element in set C is related to the unique element in M. That is, $f:C \to M$ is a function, if for each $c_i \in C$ there exists a unique $m_j \in M$ such that $f(c_i) = m_j$ [6].

The pictorial representation of the function $f:C \to M$ is as follows.

If $f:C \to M$ is a function from C to M and $f(c_i) = m_j$ then

i. m_j is known as the image of c_i.
ii. c_i is called as the pre-image of m_j.
iii. The sets C and M are called as Domain set and Co-domain set, respectively.
iv. The range set of the function is defined as

$$\text{Range Set} = R(f) = \left\{m_j/\exists c_i \text{ such that } f(c_i) = m_j\right\}$$

It is also known as the Image set of the function. Moreover, $R(f) \subseteq M$.

Examples

Example 5.1 Determine whether the following are functions or not. If functions, find the range sets.

i. Let $f:Z \to Z$ such that $f(x) = x^2$, $Z =$ Set of Integers.

ii. Let $g{:}Z \to Z$ such that $g(x) = \sqrt{x}$.
iii. Let $h{:}N \to N$ such that $h(x) = x^2 + 13$, N = Set of Natural Numbers
iv. Let $p{:}R \to R$ such that $p(x) = 2x - 16$, R = Set of Real Numbers

Solution Let Z, N, and R be the set of Integers, Natural Numbers, and Real Numbers, respectively.

i. For any integer x, it's square is also an integer and it is unique. Therefore, f is a function and it's range set is the set integers which are perfect squares. That is,

$$R(f) = \left\{x^2/\,\forall x \in Z\right\}.$$

ii. As the square root of 10 is not an integer, so, g is not a function.
iii. For any $x, y \in N \Rightarrow x^2, x + y, x^2 + 13$ are also natural numbers. Therefore, the relation $h{:}N \to N$ is a function. The smallest natural number is 1. Therefore, the range is as follows.

$$R(h) = \{x + 13/\,\forall x \in N\} = \{14, 15, 16, \ldots\}$$

iv. For any $x, y \in R \Rightarrow 2x, x - y, 2x - 16 \in R$. Therefore, p is a function and the range set is given below.

$$R(p) = \{2x - 16/\,\forall x \in R\} = R.$$

Partial Functions

A relation $f{:}C \to M$ is not a function and $H \subset C$ such that $f_H{:}H \to M$ is a function which is called as the partial function [6].

The pictorial representation of the partial function is as follows (Fig. 5.2).

The above figure depicts the partial function.

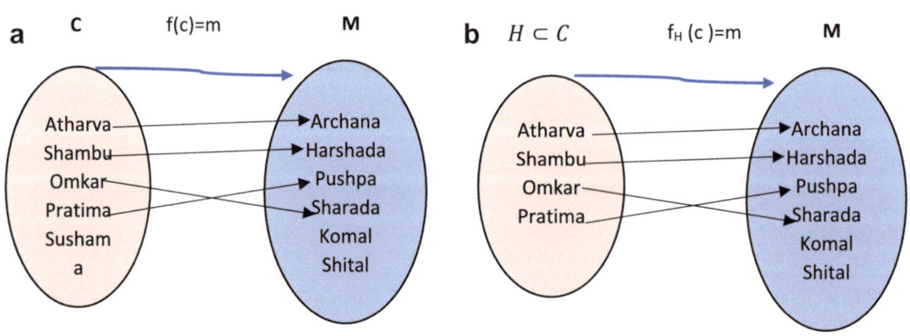

Fig. 5.2 **a** Not function, **b** partial function

Examples

1. A relation $f:R \to R$ defined as $f(x) = \sqrt{x}$ is not a function as $\sqrt{x}$ is not defined for $x < 0$ in the set of real numbers. But, $A = R^+ \cup \{0\} \subset R$ is the set of non-negative real numbers and $f:A \to R$ is the function called as the partial function.
2. A relation $p:Z \to Z$ defined as $p(m) = \frac{1}{m}$ is not a function as $\frac{1}{m}$ is not defined for $m = 0$ But, $A = Z - \{0\} \subset Z$ is the set of non-zero integers and $p:A \to Z$ is the Partial function.

5.3 Types of Functions: Experiential Learning

Consider the following special inferences of Experiment-I.

Inferences-II

In Experiment-I, we have observed the following special inferences.

1. It is possible that every child is single child of the mother in the set C. That means in C, there are no any brothers or sisters or brothers and sisters. Furthermore, a child of every mother of set M is present in C. This concept leads to the "*ONE TO ONE FUNCTION or INJECTIVE FUNCTION.*"
2. If in the set C, two brothers or two sisters or a brother and a sister are present then it leads to "*NOT INJECTIVE.*"
3. If in this relation R, every mother has a child or children which are present in the set C then this becomes a particular relation that leads to the definition of "ONTO FUNCTION."
4. There may be a mother in the set M, whose child is not in C. This type motivates us to define "*INTO FUNCTION.*"
5. If the set R satisfies (1) and (3) then R signs to the "*BIJECTIVE FUNCTION.*"
6. It is possible that the set C has only 4 children who are brothers or sisters of each other. So, each child is associated with the same mother. This concept leads to "*CONSTANT FUNCTION*" [7].

Mathematical Insights-II

5.3.1 One to One Function (Injective Function)

A function $p:C \to M$ is said to be One to One Function if $C_i \neq C_j \Rightarrow p(C_i) \neq p(C_j)$ OR $p(C_i) = p(C_j) \Rightarrow C_i = C_j$. It is also known as the Injective Function.

A function $p:C \to M$ is said to be not injective function if there exists at least two members in C such that $p(C_i) = p(C_j) \Rightarrow C_i \neq C_j$ OR $C_i \neq C_j \Rightarrow p(C_i) = p(C_j)$.

Examples

1. The first case in the inference-II is the example of the one to one function.
2. The second case in the inference-II is the example of not injective function.

5.3.2 *Onto Function (Surjective Function)*

A function $p{:}C \to M$ is said to be Onto Function if for every $M_i \in M, \exists C_j \in C$ such that $p(C_j) = M_i$. That is, every element of M has pre-image in C. Alternately, the function is onto iff $R(C) = M$. It is also called as Surjective function.

A function $p{:}C \to M$ is said to be not onto function if $R(C) \subset M$ or $\exists$ at least one $M_i \in M$, such that $p(C_j) \neq M_i, \forall j$. A function which is not onto function is known as into function.

Examples

1. The Case-3 in the inference-II is the example of onto function.
2. The Case-4 in the inference-II is the example of not onto (Into) function.

5.3.3 *Bijective Function*

A function $p{:}C \to M$ is said to be Bijective Function if it is Injective as well as Surjective function. That is, $C_i \neq C_j \Rightarrow p(C_i) \neq p(C_j)$ and $M_i \in M, \exists C_j \in C$ such that $p(C_j) = M_i$.

A function $p{:}C \to M$ is not bijective function if it is not injective or not Surjective or not both [8].

Remarks

1. A function is bijective iff $|C| = |M|$.

Examples

1. The Case-5 in the inference-II is the example of onto function.
2. The Case-2 and Case-4 in the inference-II are not bijective functions.

5.3.4 *Constant Function*

A function $p{:}C \to M$ is called Constant Function if $|R(C)| = 1$ i.e.

$$p(C_1) = p(C_2) = p(C_3) \cdots = p(C_{|C|}) = M_0.$$

The constant function has only one element in it's range set.

Examples

1. The Case-6 in the inference-II is the example of constant function.

Special Types of Functions

1. **Identity Function**: A function $p:C \to C$ is called the Identity Function if $p(C_i) = C_i$ for all i.
2. **Characteristic Function of a Set**: Let U be the Universal set, and C be the subset of U. The characteristic function $\psi_C:U \to \{0, 1\}$ defined as

$$\psi_C(x) = \left\{ \begin{array}{ll} 1; & \text{if } x \in C \\ 0; & \text{if } x \notin C \end{array} \right\}.$$

3. **Composite Function:** Let $p:C \to M$ and $h:M \to G$ be two functions. The composite function of p and h are denoted by hop and defined as hop:$C \to G$ is a function such that

$$\text{hop}(c) = h\big[p(c)\big]; \quad \forall c \in C$$

Note: The composite function hop is defined only when the range of p is a subset of the domain set of h.

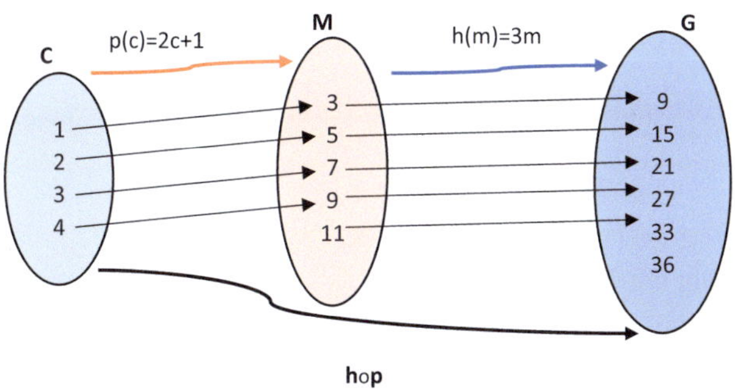

Therefore, hop:$C \to G$

$$\text{hop}(c) = h\big[p(c)\big] = h[2c + 1] = 3(2c + 1); \quad \forall c \in C$$

$\text{hop}(1) = h\big[p(1)\big] = h[3] = 9$, $\text{hop}(2) = h\big[p(2)\big] = h[5] = 15$, $\text{hop}(3) = h\big[p(3)\big] = h7 = 21$ and $\text{hop}(4) = h\big[p(4)\big] = h[9] = 27$

Note:

i. The composite function may or may not be bijective.
ii. The composite function is bijective if both the functions are bijective

4. **Inverse Function**: Let $p{:}C \rightarrow M$ be a bijective function then the function $p^{-1}{:}M \rightarrow C$ is called the Inverse function and defined as $p^{-1}(m) = c \text{ iff } p(c) = m$.

   It is also known as Invertible Function.
   Note: The inverse function is also bijective function.

Example 5.2 Give examples of the following types of functions by diagrams/ pictorial representations.

(a) Injective (1–1) but not onto
(b) onto but not 1–1
(c) Neither 1–1 nor onto
(d) 1–1 and onto

Solution

1. **Pictorial Representations**

   (a) 1–1 but not onto

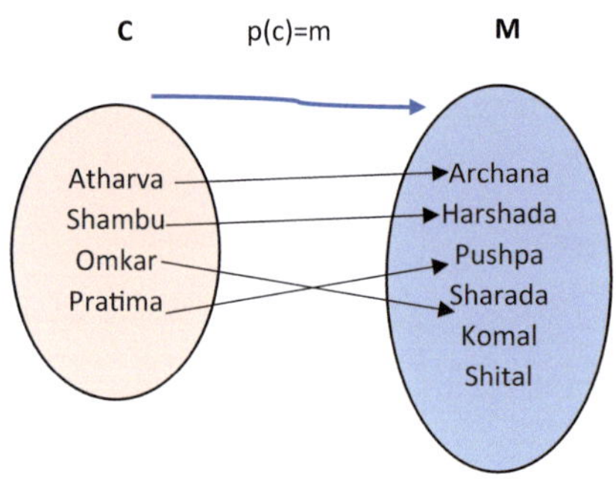

   (b) Onto but not 1–1

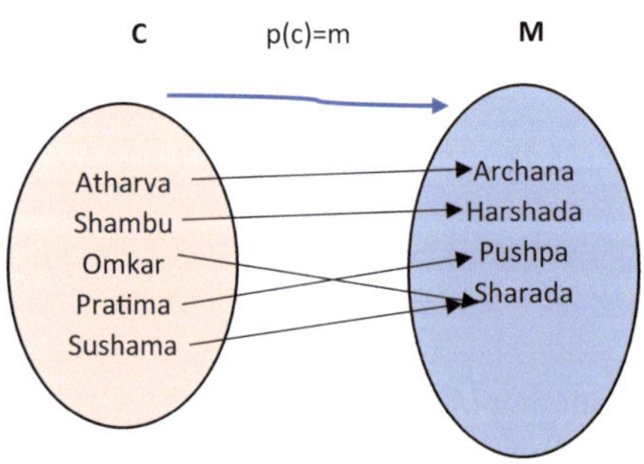

(c) Neither 1–1 nor onto

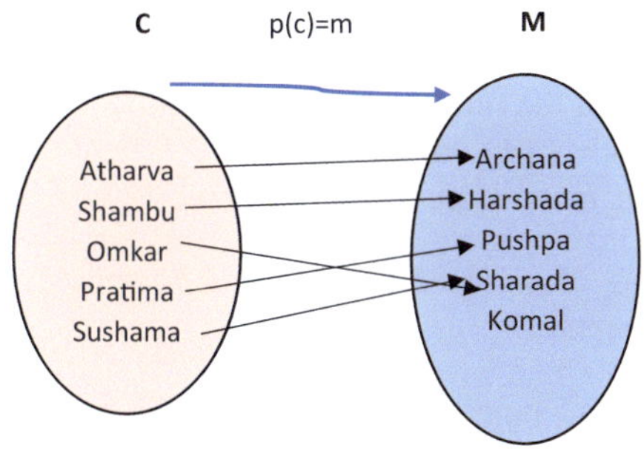

(d) Onto and 1–1

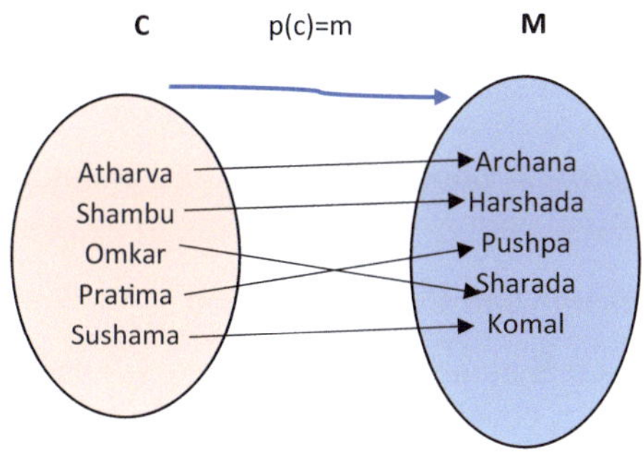

2. **Mathematical Functions**

a. 1–1 but not onto: Let $p{:}N \to N$ be a function on the set of Natural numbers N such that $p(n) = 2n$ for all n in N. So, the function is 1–1.

   Furthermore, the range set of p is $R(p) = \{2, 4, 6, 8 \ldots, 2n, \ldots\}$. The odd numbers do not have pre-image in the domain set. Therefore, the function p is injective but not onto.

b. Onto but not 1–1: Let $p{:}\{-2, -1, 0, 1, 2\} \to \{0, 1, 4\}$ be a function such that $p(n) = n^2$ for all n in N. The square of 1 and $-$ 1 is 1, that is 1 and $-$ 1 are related to 1. Therefore, p is not injective function. But, the function p is surjective as every element in the codomain set has pre-image in the domain set. Hence the result.

c. Neither 1–1 nor onto: Let $p{:}\{-2, -1, 0, 1, 2\} \to \{0, 1, 4, 9\}$ be a function such that $p(n) = n^2$ for all n in N. The function p is neither 1–1 nor onto.

d. Injective and Surjective: Let $p{:}R \to R$ be a function on the set of real numbers such that $p(n) = 3n - 1$. This function p is injective as well as surjective.

Theorem 5.1 *If $p{:}C \to M$ is bijective function then p^{-1} is unique.*

Proof Suppose p^{-1} is not unique, so there are two inverse functions say h and k. Let C_1 and C_2 be any two elements in C then there exists m in M such that

$$p^{-1}(m) = C_1 \Rightarrow h(m) = C_1 \Rightarrow p(C_1) = m$$
$$p^{-1}(m) = C_2 \Rightarrow k(m) = C_2 \Rightarrow p(C_2) = m$$

$\Rightarrow p(C_1) = p(C_2)$, but the function p is 1–1 function.
$\therefore\ C_1 = C_2 \Rightarrow h(m) = k(m), \forall m \in M$.
Therefore, $h = k$ i.e. h and k are equal functions.

Theorem 5.2 *If $p{:}C \to M$ and $h{:}M \to G$ are two bijective functions then hop:$C \to G$ is also a bijective function.*

Proof Let $C_1, C_2 \in C$ and suppose $\text{hop}(C_1) = \text{hop}(C_2)$

$$h[p(C_1)] = h[p(C_2)] \Rightarrow p(C_1) = p(C_2) \Rightarrow C_1 = C_2.$$

Therefore, hop is 1–1 function?

Let $g \in G$ be any element. $\exists m \in M$ such that $h(m) = g$ and $\exists c \in C$ *suchthat* $p(c) = m$.

$\Rightarrow g = h(m) = h[p(c)] = \text{hop}(c)$. Therefore, *hop* is onto function. Hence, *hop* is bijective function.

Definition: Floor Function and Ceiling Function

The Floor function of a real number p is denoted by $\lfloor p \rfloor$ and defined as the largest integer that is less than or equal to p. The Ceiling function of a real number p is denoted by $\lceil p \rceil$ and defined as the smallest integer that is greater than or equal to p [8].

For example, $\left\lceil \frac{1}{3} \right\rceil = 1$, $\left\lfloor \frac{1}{3} \right\rfloor = 0$, $\left\lceil \frac{-1}{2} \right\rceil = 0$, $\left\lfloor \frac{-1}{2} \right\rfloor = -1$, $\lceil 10 \rceil = 10 = \lfloor 10 \rfloor$.

Properties of the Floor and Ceiling Functions are given below. Here n is an integer.

- $\lceil p \rceil = n \text{ iff } n - 1 < p \le n$,
- $\lceil p \rceil = n \text{ iff } p \le n < p + 1$,
- $\lfloor p \rfloor = n \text{ iff } n \le p < n + 1$,
- $\lfloor p \rfloor = n \text{ iff } p - 1 < n \le p$,
- $p - 1 < \lfloor p \rfloor \le p \le \lceil p \rceil \le p + 1$,
- $\lceil -p \rceil = -\lfloor p \rfloor$,
- $\lfloor -p \rfloor = -\lceil p \rceil$,
- $\lceil p + n \rceil = \lceil p \rceil + n$,
- $\lfloor p + n \rfloor = \lfloor p \rfloor + n$.

Example 5.3 Prove or disprove that $\lceil z + w \rceil = \lceil z \rceil + \lceil w \rceil$ for all real numbers z and w.

Solution This type statement is true in many cases, but it is false for the ceiling function. For a counter example, take $z = \frac{1}{3}$ and $w = \frac{2}{3}$.Therefore,

$$\lceil z \rceil = 1, \lceil w \rceil = 1, \text{ but } \lceil z + w \rceil = \lceil 1 \rceil = 1.$$

Example 5.4 Prove or disprove that $\lfloor z + w \rfloor = \lfloor z \rfloor + \lfloor w \rfloor$ for all real numbers z and w.

Solution This result is not true for all real numbers. Take $z = \frac{1}{2}$ and $w = \frac{1}{2}$. Therefore, $\lfloor z \rfloor = 0$, $\lfloor w \rfloor = 0$, but $\lfloor z + w \rfloor = \lfloor 1 \rfloor = 1$.

Example 5.5 Let C and M be the set of positive real numbers such that $m > 13, \forall m \in M$ and $p{:}C \to M$ and $h{:}M \to C$ are defined as $p(c) = c^3 + 13$ and $h(m) = \sqrt[3]{m - 13}$. Prove that p is bijective function and h is the inverse of p.

Solution Let $p(c_1) = {c_1}^3 + 13$ and $p(c_2) = {c_2}^3 + 13$. Suppose $p(c_1) = p(c_2)$.
$\Rightarrow {c_1}^3 + 13 = {c_2}^3 + 13 \Rightarrow c_1 = c_2 \Rightarrow p$ is 1–1 function.
For any $m \in M$, $\exists c$ such that $p(c) = m \Rightarrow c^3 + 13 = m \Rightarrow c = \sqrt[3]{m - 13} \in M$.
Therefore, p is onto function. Thus, p is bijective function. Moreover, $p^{-1} = h \Rightarrow h$ is also bijective function.

Example 5.6 Determine whether the function p is bijective or not if $p{:}Z \to Z$ such that

$$\begin{aligned} p(n) &= \frac{w}{2}; \quad \text{if } w \text{ is even} \\ &= \frac{w-1}{2}; \quad \text{if } w \text{ is odd} \end{aligned}$$

where Z is the set of integers.

Solution Let n and m be two integers.

Case 1: If n and m are odd integers.

$$p(n) = p(m) \Rightarrow \frac{n-1}{2} = \frac{m-1}{2} \Rightarrow n = m.$$

Case 2: If n and m are even integers.

$$p(n) = p(m) \Rightarrow \frac{n}{2} = \frac{m}{2} \Rightarrow n = m.$$

Case 3: If n is even and m is odd.

$$p(n) = p(m) \Rightarrow \frac{n}{2} = \frac{m-1}{2} \Rightarrow n = m - 1 \Rightarrow n - m = 1 \Rightarrow n \neq m.$$

Therefore, the function p is not 1–1 function. Thus, p is not bijective function.

Example 5.7 Let p and h be the functions on the set of real numbers such that $p(x) = ax^2 + b$ and $h(x) = cx^2 + d$, where a, b, c, d are real numbers. Determine for which values of constants $\mathrm{poh}(x) = \mathrm{hop}(x)$.

Solution We have

$$\text{poh}(x) = \text{hop}(x) \Rightarrow p[h(x)] = h[p(x)]$$
$$p[cx^2 + d] = h[ax^2 + b]$$
$$a(cx^2 + d)^2 + b = c(ax^2 + b)^2 + d$$
$$a(c^2x^4 + 2cdx^2 + d^2) + b = c(a^2x^4 + 2abx^2 + b^2) + d$$

Coefficients of x^4 in LHS = Coefficients of x^4 in RHS and

Coefficients of x^4 in LHS = Coefficients of x^4 in RHS

$$\Rightarrow ac^2 = ca^2 \text{ and } 2acd = 2abc$$
$$\Rightarrow c = a \text{ and } d = b$$

Example 5.8 Show that the functions such that $h(x, y) = x + y$ and $p(x, y) = xy$ are onto but not one–one.

Solution

(i) Every element of the N can be expressed as the sum and the product of the two elements of N. Therefore, the functions h and p are onto functions.

(ii) Let (x_1, y_1) and $(x_2, y_2) \in N \times N$

$h(x_1, y_1) = h(x_2, y_2) \Rightarrow x_1 + y_1 = x_2 + y_2 \Rightarrow$ may be $x_1 \neq x_2$ and $y_1 \neq y_2$
e.g., $x_1 = 10, x_2 = 13$ and $y_1 = 6, y_2 = 3$.
Therefore, h is not one–one function.
Now, $p(x_1, y_1) = p(x_2, y_2) \Rightarrow x_1y_1 = x_2y_2 \Rightarrow$ may be $x_1 \neq x_2$ and $y_1 \neq y_2$.
e.g., $x_1 = 10, x_2 = 20$ and $y_1 = 6, y_2 = 3$.
Therefore, p is not one–one function.

5.4 Infinite Sets and Countability: Experiential Learning

(A) Brainstorming Experiment-III

Let H be the set of all existing words in English language. We can count these words, but there are many English words. If H_1 is the possible number of words then in all circumstances, H_1 is the proper subset of H as there could be a word which is not in H.

Let's consider the following inferences.

(B) Inferences-III

In Experiment-I, we have observed the following probable special inference.

1. For any H_1, it is a subset of H. This concept leads to the definition of Infinite Set.

(C) Mathematical Insights-III

5.4.1 Infinite Set

A set H is said to be an infinite set if there exists an injective function $p:H \rightarrow H$ such that $p(H) \subset H$, i.e., $p(H)$ is a proper subset of H. Otherwise, it is finite set [2].

Properties of Infinite Set

1. If H is an infinite set then $H \times H$ and $P(H)$ are infinite sets.
2. If $H \neq \phi$ and $K \neq \phi$ and either H or K is an infinite set then $H \times K$ is an infinite set.
3. If either H or K is an infinite set then $H \cup K$ is an infinite set.
4. If H is an infinite set and $H \subseteq K$ then K is also an infinite set.
5. If H and K are infinite sets then $H \cap K$ may or may not be an infinite set.
6. A subset of an infinite set may or may not be infinite.

Examples:

Example (1) Let $p:N \rightarrow N$ such that $p(n) = 2n, \forall n \in N$, where N is the set of natural numbers. The range set $p(N) = \{2n/n \in N\} \subset N$. Therefore, N is an infinite set.

5.4.2 Countable Sets

We know that, the set of natural numbers are used to count the objects. Let's use the flavor of set of natural numbers with the function and define the countable sets.

Definition: An infinite set H is said to be countable set, if there exist a bijective function $p:N \rightarrow H$. Therefore, the set H can be written as

$$p(H) = \{p(1), p(2), p(3), \ldots, p(n), \ldots\}.$$

Alternately and more precisely, the definition of countable set is given below.

An infinite set H is said to be countable set if there exist a bijective function $p:N_1 \rightarrow H$, where, $N_1 \subseteq N$.

Notes

i. A countably infinite set is known as a denumerable set.
ii. Every finite set is countable.
iii. The empty set ϕ is countable.
iv. If H and K are countable sets then $H \cup K$ and $H \cap K$ are countable sets.
v. The countable union of the countable sets is countable.

Examples

Example 5.9 Prove that the set of integers is countable.

Solution Define $p{:}N \rightarrow Z$ as

$$p(n) = \frac{n+1}{2}, \quad \text{if}\, n = 1, 3, 5, \ldots$$
$$= \frac{-n}{2}, \quad \text{if}\, n = 0, 2, 4, \ldots$$

The function p is bijective function. Hence the result.

Example 5.10 Prove that the set of rational numbers is countable.

Solution Let's define the function $p{:}Q \rightarrow Z \times N$, where Q, Z, and N are the set of rational numbers, integers, and natural numbers, respectively, as

$$p\left(\frac{m}{n}\right) = (m, n), \forall \frac{m}{n} \in Q \quad \text{where, } (m, n) = 1.$$

The function p is injective. The Cartesian product of countable sets is countable. Hence the proof.

Example 5.11 Prove that the set of real numbers in (0, 1) is not countable.

Solution Assume that the set is countable. Any real number in (0, 1) can be written in a unique decimal without an infinite string of 9's at the end. i.e., 0.13499999 can be represented as 0.13400.

Let the infinite countable sequence be given as

$$1 \rightarrow h_1 = 0 \ldots x_{11}\, x_{12}\, x_{13} \;\ldots$$
$$2 \rightarrow h_2 = 0 \ldots x_{21}\, x_{22}\, x_{23} \;\ldots$$
$$3 \rightarrow h_3 = 0 \ldots x_{31}\, x_{32}\, x_{33} \;\ldots$$
$$\vdots$$
$$n \rightarrow h_n = 0 \ldots x_{n1}\, x_{n2}\, x_{n3} \;\ldots$$
$$\vdots$$

Now, construct a new number in (0, 1) as $x = 0.b_1\, b_2\, b_3\, b_4 \ldots$ such that

$$b_i = 0 \quad \text{if}\, x_{ii} \neq 0$$
$$b_i = 1 \quad \text{if}\, x_{ii} = 0$$
$$b_i \neq x_{ii}, \qquad \forall i$$

Hence, $x \neq h_i, \forall i$

Therefore, x is not in the list $\{h_1, h_2, h_3, h_4, \ldots, h_n, \ldots\}$. Thus, $x \in (0, 1)$ but not in the list $\{h_1, h_2, h_3, h_4, \ldots, h_n, \ldots\}$. This is contradiction to assumption that the set is countable. Hence the proof.

Example 5.12 Prove that the set of real numbers R is not countable.

Solution We know that if $H \subseteq M$ and H is not countable then M is also not countable. We have $(0, 1) \subset R$ and the set of real numbers in (0, 1) is not countable. Hence, the set of real numbers is not countable.

5.5 Pigeonhole Principle: Experiential Learning

(A) Brainstorming Experiment-IV

Suppose, there are 10 different books of Discrete Mathematics in the departmental library and 9 students are studying this course. How could we distribute all these books (10 books) among 09 students? How could we distribute 10 books among 03 students?

Let's consider the following inferences.

(B) Inferences-IV

In Experiment-II, we have observed the following special inferences.

1. There is a student who receives two books of Discrete Mathematics.
2. There is no one to one relation between Books and Students. That is, the many to one function which leads to the concept of the Pigeonhole Principle.
3. In the second case, give 03 books to each student and one book to any one of them. Therefore, one student receives 04 books and other two students receive 03 books each. This case demonstrates the Generalized or Extended Pigeonhole Principle.

(C) Mathematical Insights-IV

Pigeonhole Principle

Statement: If there are $p + 1$ Pigeons and only p pigeonholes then two pigeons will share the same hole. Alternately, for any positive integer p, if $p + 1$ objects are placed into p boxes then at least one box containing two or more objects.

This principle is also derived from the concept of the function. If there are two sets P and H such that $|P| > |H|$ and $p{:}P \to H$ is a function then at least two elements of the set P have the same image in H. That is, the function p is not one–one function from P to H [2, 4].

Example 1: If there are 366 students in the department of Mathematics in 2025 then there must be at least two students having the same birthdate. (There are only 365 different dates in 2025).

The Generalized Pigeonhole Principle: If p objects are placed into h boxes and $p > h$ then there is at least one box containing at least $\left\lceil \frac{p}{h} \right\rceil$ objects, where $\lceil\ \rceil$ is the ceiling function. OR

If p objects are placed into h boxes and $p > h$ then there is at least one box containing at least $\left\lfloor \frac{p-1}{h} \right\rfloor + 1$ objects, where $\lfloor\ \rfloor$ is the floor function.

Proof Assume that each box contains at least $\left\lfloor \frac{p-1}{h} \right\rfloor$ objects. Therefore, the maximum number of objects in all the boxes $= h\left\lfloor \frac{p-1}{h} \right\rfloor \leq h\left(\frac{p-1}{h}\right)$. But, $\left\lfloor \frac{p-1}{h} \right\rfloor \leq \left(\frac{p-1}{h}\right)$. So, the maximum number of objects in all boxes is $\leq p - 1$, which is the contradiction as the total number of objects is p.

Example 5.13 shows that if 13 colors are used to paint 100 bicycles, then at least 8 bicycles are of the same color.

Solution By the generalized pigeonhole principle, at least $\left\lceil \frac{p}{h} \right\rceil = \left\lceil \frac{100}{13} \right\rceil = 8$ bicycles will have the same color.

Example 5.14 If there are 13 boys and 15 girls in a class and these students are lined up in a row then show that at least two girls will be next to each other.

Solution This example is the analogy of placing 15 objects in 13 boxes. Here, the number of objects are more than the number of boxes. Therefore, by the pigeonhole principle, there must be at least two girls nest to each other.

Example 5.15 Find the minimum number of students in the class such that 5 of them are born on the same month.

Solution We have, $h = 12$ and

$\left\lceil \frac{p}{h} \right\rceil = 5 \Rightarrow \left\lceil \frac{p}{12} \right\rceil = 5 \Rightarrow \lceil p \rceil = 60$. This is the maximum value of p. The minimum value is $p = 12 \times 4 + 5 = 53$.

Example 5.16 In a class of 64 students, how many of them are born on the same month?

Solution Here, $p = 64$, $h = 12$,

Therefore, the number of students born on the same month is

$$= \left\lfloor \frac{p-1}{h} \right\rfloor + 1 = \left\lfloor \frac{64-1}{12} \right\rfloor + 1 = 6$$

5.6 Discrete Numeric Function: Experiential Learning

Discrete Numeric Function

A function whose domain set is countable set and the codomain set is a subset of the uncountable set is called a discrete numeric function. It is also known as a sequence. It is a function $p{:}N \cup \{0\} \rightarrow R$. The values of function p at 0, 1, 2, 3, … are, respectively, denoted by $p(0)$, $p(1)$, $p(2)$, $p(3)$, … [4].

The discrete numeric function p can be written as

$$p = \{p(0), p(1), p(2), p(3), \ldots\} = \{p_0, p_1, p_2, p_3, \ldots\}.$$

Examples

Example 5.17 Let

$$\begin{aligned} p_r &= 3^r \quad \text{if } 0 \le r \le 4 \\ &= r \quad \text{if } 4 < r \end{aligned}$$

The discrete numeric function can be written as $p = \{1, 3, 9, 27, 81, 5, 6, 7, 8 \ldots\}$.

Example 5.18 Let $p_r = 0.1 \times r; \forall r$. Therefore, $p = \{0, 0.1, 0.2, 0.3, \ldots\}$.

Properties of Discrete Numeric Function

Let p and q be two discrete numeric functions and a and b are constants.

1. Addition: $h = p + q$ is a discrete numeric function defined as $h_r = p_r + q_r$.
2. Multiplication: $h = p \times q$ is a discrete numeric function defined as $h_r = p_r \times q_r$.
3. Scaling: $h = a \times p$ is a discrete numeric function defined as $h_r = a \times p_r$.
4. Linearity: $h = ap \times bq$ is a discrete numeric function defined as $h_r = ap_r + bq_r$.
5. Convolution of Two Functions: $h = p * q$ is a discrete numeric function defined as

$$h_r = \sum_{n=0}^{r} p_n q_{r-n} = p_0 q_r + p_1 q_{r-1} + p_2 q_{r-2} + p_3 q_{r-3} + \cdots + p_r q_0.$$

Finite Differences of Discrete Numeric Function

1. **Shift Function**: The shift function of the discrete numeric function is defined as

$$E[p_r] = p_{r+1}.$$

In general, $E^n[p_r] = p_{r+n}$ and $E^{-n}[p_r] = p_{r-n}$.

2. **Forward Difference**: The forward difference of the discrete numeric function is defined as

$$\Delta p_r = p_{r+1} - p_r;\ r \geq 0.$$

3. **Backward Difference**: The backward difference of the discrete numeric function is defined as

$$\nabla p_r = p_r - p_{r-1};\ r \geq 1.$$

Examples

Example 5.19 If $p_r = 5^r$; $r \geq 0$ then find

(a) $E^4[p_r]$
(b) $\Delta^2 p_r$

Solution

(a) We have $E^4[p_r] = p_{r+4} = 5^{r+4} = 625 \times 5^r$
(b) We have,

$$\begin{aligned}\Delta^2 p_r &= \Delta(\Delta p_r)\\ &= \Delta(p_{r+1} - p_r)\\ &= \Delta p_{r+1} - \Delta p_r\\ &= p_{r+2} - p_{r+1} - (p_{r+1} - p_r)\end{aligned}$$

$$\begin{aligned}\Delta^2 p_r &= p_{r+2} - p_{r+1} - (p_{r+1} - p_r)\\ &= p_{r+2} - 2p_{r+1} + p_r\end{aligned}$$

5.7 Case Studies and Projects

The following case studies/projects are given for improving and developing deep understanding of the concepts.

1. Design an examples from daily life that follow the function and its all types.
2. Election voting system and associated functions.
3. Applications of countable and uncountable sets. Design your own 10 examples of countable sets and 10 examples of uncountable sets.
4. Make a report on applications of Pigeon hole principle in real life (Five Applications)
5. Importance of discrete numeric functions and applications.

Exercise

1. Design an examples of the following
   i. Relation but not function
   ii. Function but not 1–1 function
   iii. Function but not onto function
   iv. 1–1 function and onto function
   v. 1–1 function but not onto
   vi. Onto but not 1–1 function
   vii. Into function but not 1–1 function
2. Prove that the composition of bijective functions is a bijective function.
3. Prove that $f, g: N \times N \rightarrow N$ as $f(x, y) = x + y$ and $g(x, y) = xy$ are onto functions but not 1–1 functions.
4. Prove that all numbers lie on the positive x-axis are uncountable.
5. The set $H = \{a + ib/\ a$ and b are integers$\}$ is countable.
6. Show that 7 colors are used to paint 50 bicycles, then at least 8 bicycles are of the same color.
7. If $H = \{1, 2, 3, ..., 100\}$, show that any function from H to H which is 1–1 must be onto, and conversely.
8. List all possible functions from the set $H = \{p, q, r, s\}$ to the set $M = \{1, 2, 3\}$and specify the types of the function with proof.
9. Prove that $N \times N$ is countable set, where N is the set of natural numbers.
10. In a group of 100 people, how many of them will have the same birth day.

References

1. Aigner M (2023) Discrete mathematics. American Mathematical Society
2. Erciyes K (2021) Discrete mathematics and graph theory. Springer, Cham. https://doi.org/10.1007/978-3-030-61115-6
3. Rosen K, Krithivasan K (2021) Discrete mathematics and its applications. Mc Graw Hill India. ISBN-978930727353
4. Jiri M, Jaroslav N (2011) Invitation to discrete mathematics, 2nd edn. Oxford University Press. ISBN-9780198570424
5. Peter J (2000) Sets, logic and categories. Springer Publication. ISBN-9781852330569
6. Tremblay JP, Manohar R (2023) Discrete mathematical structures with applications to computer science. McGraw Hill Education. ISBN-9780074631133
7. Baker KA, Fishburn PC, Roberts FS (1971) Partial orders of dimension 2. Networks 2(1):11–28. https://doi.org/10.1002/net.3230020103
8. Birkhoff G (1948). Lattice theory, rev edn. American Mathematical Society

Chapter 6
Logic and Proofs

> *"Logic and mathematics are nothing but specialised linguistic structures"*
> —Jean Piaget

Abstract This chapter illustrates how Logic and Mathematics are specialized specialized linguistic structures. Mathematics is the language of science and technology. The logic assists us in deciding the validity of the arguments. The truth and falsity of the arguments are directly validated by mathematical theory. This chapter explains the concepts of different types of propositions, truth tables, connectives, inferences, properties, and quantifiers through the brainstorming activities or experiments. This chapter explains truth tables and various logical operations such as Conjunction, Disjunction, Conditional and biconditional statements through an innovative approach to the operations of set theory. This analogy enhances the understanding of the concepts and their applications in different ways. Furthermore, it elucidates the concepts of tautology and contradiction with suitable examples. It also helps to improve the mathematical and logical thinking of the readers through the concepts of Algebra of Propositions, quantifiers and its applications. Furthermore, this chapter explores the hypothesis of the concepts of conditional and biconditional statements. The chapter concludes with case studies and exercises aimed at connecting theoretical relational structures to real-world applications across various disciplines.

Keywords Truth table · Logical statements · Algebra of propositions · Logical equivalence · Quantifiers

6.1 Introduction

Mathematics plays an imperious role in increasing the gray matter of the brain. The mathematical thinking is more precise and accurate as compared with others. It aids in fast and more pragmatic decision-making. The average aptitude of mathematics students is better than other students. Logical thinking is essential for understanding

© The Author(s), under exclusive license to Springer Nature Singapore Pte Ltd. 2026

H. R. Bhapkar and P. N. Mahalle, *Discrete Mathematics: An Experiential Learning Perspective*, Studies in Autonomic, Data-driven and Industrial Computing,
https://doi.org/10.1007/978-981-95-7210-6_6

difficult concepts and topics. The logic assists us in deciding the validity of the arguments. The truth and falsity of the arguments are directly validated by the mathematical theory. This chapter explains the concepts of different types of propositions, truth tables, connectives, inferences, properties, and quantifiers [1, 2].

6.2 Propositions and Compound Statements: Experiential Learning

Let's consider the following brainstorming Experiment-I for understanding the concept of the propositions and compound statements

(A) **Brainstorming Experiment-I**

Many students are playing a game of telling the non-related consecutive sentences. That is, there should not be any relation between two sentences. Suppose, the first sentence is "Mr. Atharva is a clever boy." So, the second sentence must be free from the words Atharva, clever and boy. The second sentence may be "Delhi is the capital of India or Today's weather is glossy." Moreover, if Mr. Atharva is a clever boy then Delhi is the capital of India. This sentence has no meaning in Mathematics. The connected sentences must be logically significant.

In real life, we come across many situations where the following statements are used.

i. Mr. Atharva is a clever boy.
ii. Mr. Omkar is honest, and he is very sincere.
iii. If Atharva is a hardworking boy, then he will get excellent percentile in the examination.
iv. Omkar is very sincere in his studies iff he will get a good percentage in his B. Tech. examination.

Let us understand the inferences in different ways.

(B) **Inferences-I**

In Experiment-I, we have observed the following special inferences.

i. The statement (i) is a single sentence. We call it as a **proposition or statement**.
ii. The statement (ii) involves two sentences that are connected by "and." This is the **compound proposition**.
iii. The statement (iii) is a compound proposition connected by "If... then....". This is known as the **Conditional proposition**.
iv. The statement (iv) is compound statement having "If and only if" connectives. This is called as **Biconditional proposition**.

(C) Mathematical Insights-I

Consider the following mathematical insights that are obtained from Experiment-I.

Propositions or Statements

A declarative sentence which is either true or false but not both, is called a Proposition or Statement. The truth and the falsity of a statement is called it's truth value.

"T" or "1" and "F" or "0" are used to denote the truth value and false value [3].

Examples

1. There are 25 days in a month of August.
2. $13 + 5 = 18$
3. $X + 5 = 17$
4. There are seven days in a week.
5. There will be rain tomorrow in Pune

Examples (2) and (4) are true sentences and example (1) is false. So, the sentences given in examples (1), (2), and (4) are statements of propositions. But, in example (3), the truth value of an equation depends upon the value of X. If $X = 12$, then the equation is true, otherwise it is false. Therefore, the sentence (3) is not a statement.

We can't predict the rain, but it will be decided by tomorrow and it will be either true or false. Hence, example (5) is a statement.

In broad spectrum, there are only two laws of the formal logic. The law of contradiction means the statement cannot be true and false. The law of omitted middle means there is no any possible stage between True and False.

Compound Statements: A statement or proposition which is formed by using two or more primary statements, is called a compound statement.

The statements given in the experiments (ii), (iii), and (iv) are compound statements.

6.3 Basic Logical Operations: Experiential Learning

From Experiment-I, the inferences (ii), (iii), and (iv) are used the logical connectives for connecting two propositions together.

A statement is called an atomic or primary or primitive statement if it can't be further divided into simple sentences. There are many connectives in languages (i.e., NOT, AND, OR, BUT, …), but these connectives are flexible in their meanings, and leads to imprecise and unclear interpretations. However, Mathematics is a language of science and technology having very precise meaning and interpretations [1, 4].

Let's consider the following special connectives in mathematics having the precise meaning depicted in Table 6.1.

e.g., let p: Atharva is studying in 10th standard

Table 6.1 Connectives in mathematics

S. No	Name of connectives	Symbol or notation
1	Negation	$\neg$ or ~
2	And or Meet	$\wedge$
3	Or (Join)	$\vee$
4	If… then	$\rightarrow$
5	If and only if	$\leftrightarrow$

and q: Atharva is learning basic algebra.

Therefore, the compound statements by using the connectives are as follows.

- Atharva is studying in 10th standard and Atharva is learning basic algebra, i.e., $p \wedge q$.
- Atharva is studying in 10th standard or Atharva is learning basic algebra, i.e., $p \vee q$.
- If Atharva is studying in 10th standard then Atharva is learning basic algebra, i.e., $p \rightarrow q$.
- Atharva is studying in 10th standard iff Atharva is learning basic algebra, i.e., $p \leftrightarrow q$.

6.4 Propositions and Truth Tables

The truth tables are formed by using the truth value of the statement formulae. Let's consider the following truth tables of the various statements.

6.4.1 Negation

The negation of a statement h is designed by writing "It is false that" before the statement h or putting the word "not" in the statement at the proper place. Notation: $\sim h$. The negation is an unary connective as it requires only one statement [6].

The truth table of negation is depicted in Table 6.2.

Table 6.2 Truth table of negation

h	$\sim h$
T	F
F	T

Example 6.1

Sr. No.	Statement h	Statement ~ h
1.	M is the capital of T state	M is not the capital of T state It is false that M is the capital of T state It is not the case that M is the capital of T state
2.	$x > 10$	$x \leq 10$ ~ p: It is false that $x > 10$
3.	$p + s = 25$	$p + s \neq 25$

6.4.2 Conjunction

A conjunction $h \wedge m$ is designed by joining two primary statements h and m, by using the connective "AND." The statement $h \wedge m$ is T if both h and m are true, otherwise false.

The truth table of $h \wedge m$ is depicted in Table 6.3.

Examples Example 6.2: Let h: p be an element in B, m: p be an element in M.

There are the following possibilities for $B \cap M$.

1. If $p \in B$ and $p \in M$ then $p \in B \cap M$.
2. If $p \in B$ and $p \notin M$ then $p \notin B \cap M$.
3. If $p \notin B$ and $p \in M$ then $p \notin B \cap M$.
4. If $p \notin B$ and $p \notin M$ then $p \notin B \cap M$.

These four cases are depicted in Fig. 6.1.

Consider the following truth table that represents the rule of conjunction by using the above properties of the union two sets. In this, H = set of integers in $-\infty < x \leq 10$ and K = set of integers in $0 \leq x < \infty$.

Table 6.3 Truth table of negation

h	M	$h \wedge m$
T	T	T
T	F	F
F	T	F
F	F	F

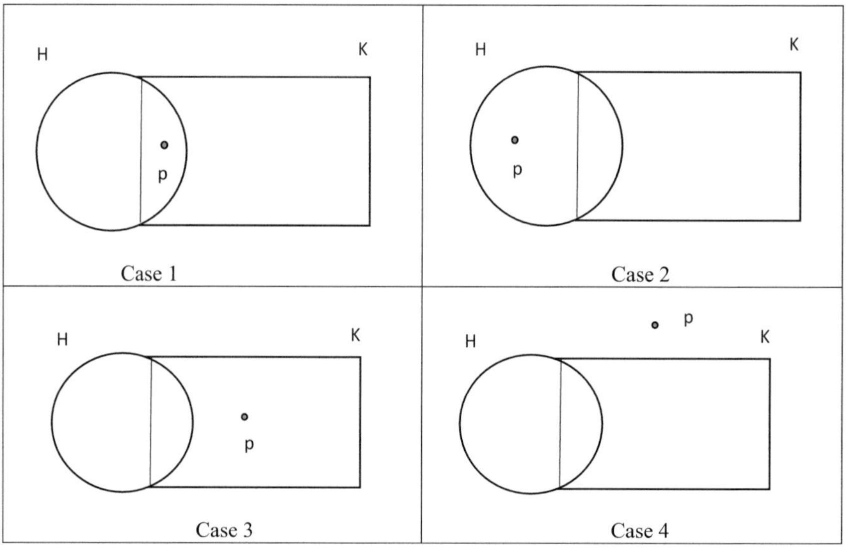

Fig. 6.1 Set theory approach for conjunction

Sr. No.	h	m	Conjunction statement	$h \wedge m$
1.	$x \in H$ **True** (i.e., 5)	$x \in K$ **True** (i.e., 5)	$x \in H$ and $x \in K$ **True** (i.e., 5)	**T**
2.	$x \in H$ **True** (i.e., -13)	$x \in K$ **False** (i.e., -13)	$x \in H$ and $x \in K$ **False** (i.e., -13)	**F**
3.	$x \in H$ **False** (i.e., 15)	$x \in K$ **True** (i.e., 15)	$x \in H$ and $x \in K$ **False** (i.e., 15)	**F**
4.	$x \in H$ **False** (i.e., 2.5)	$x \in K$ **False** (i.e., 2.5)	$x \in H$ and $x \in K$ **False** (i.e., 2.5)	**F**

Example 6.3

Sr. No.	h	m	Conjunction statement	$h \wedge m$
1.	$10 > 7$ **(T)**	7 is a prime number **(T)**	$10 > 7$ **and** 7 is a prime number	**T**
2.	11,025 is the square of 105 **(T)**	4 is a prime number **(F)**	11,025 is the square of 105 **and** 4 is a prime number	**F**

(continued)

(continued)

Sr. No.	h	m	Conjunction statement	$h \wedge m$
3.	5 is a square root of 10 **(F)**	11 is a prime number **(T)**	5 is a square root of 10 **and** 11 is a prime number	**F**
4.	$3 + 10 = 15$ **(F)**	$15 + 10 = 20$ **(F)**	$3 + 10 = 15$ and $15 + 10 = 20$	**F**

6.4.3 Disjunction

A disjunction ($h \vee m$) of two primary statements h and m is designed by using the connective "OR." The statement $h \vee m$ is F if both h and m are false, otherwise it is true [6].

The truth table of $h \vee m$ is depicted in Table 6.4.

Example 6.4

Sr. No.	h	m	Disjunction statement	$h \vee m$
1.	$10 > 7$ **(T)**	113 is a prime number **(T)**	$10 > 7$ **or** 113 is a prime number	**T**
2.	11,025 is a square of 105 **(T)**	4 is a prime number **(F)**	11,025 is the square of 105 **or** 4 is a prime number	**T**
3.	5 is a square root of 10 **(F)**	11 is a prime number **(T)**	5 is a square root of 10 **or** 11 is a prime number	**T**
4.	$3 + 10 = 15$ **(F)**	$15 + 10 = 20$ **(F)**	$3 + 10 = 15$ **or** $15 + 10 = 20$	**F**

Example 6.5 Let h: p be an element in B, m: p be an element in M.

There are the following cases.

(1) If $p \in B$ and $p \in M$ then $p \in B \cup M$.

Table 6.4 Truth table of disjunction

h	m	$h \vee m$
T	T	T
T	F	T
F	T	T
F	F	F

(2) If $p \in B$ or $p \notin M$ then $p \in B \cup M$.
(3) If $p \notin B$ or $p \in M$ then $p \in B \cup M$.
(4) If $p \notin B$ or $p \notin M$ then $p \notin B \cup M$.

These four cases are depicted in Fig. 6.2.

Consider the following truth table that represents the rule of disjunctions by using the above properties of the union two sets. In this, H = set of integers x in $-\infty < x \leq 10$ and K = set of integers such that $0 \leq x < \infty$.

Sr. No.	h	m	Disjunction Statement	$h \vee m$
1	$x \in H$ **True** (i.e., 1)	$x \in K$ **True** (i.e., 1)	$x \in H$ or $x \in K$ **True** (i.e., 1)	**T**
2	$x \in H$ **True** (i.e., -100)	$x \in K$ **False** (i.e., -100)	$x \in H$ or $x \in K$ **False** (i.e., -100)	**T**
3	$x \in H$ **False** (i.e., 150)	$x \in K$ **True** (i.e., 150)	$x \in H$ or $x \in K$ **False** (i.e., 150)	**T**
4	$x \in H$ **False** (i.e., 2.55)	$x \in K$ **False** (i.e., 2.55)	$x \in H$ or $x \in K$ **False** (i.e., 2.55)	**F**

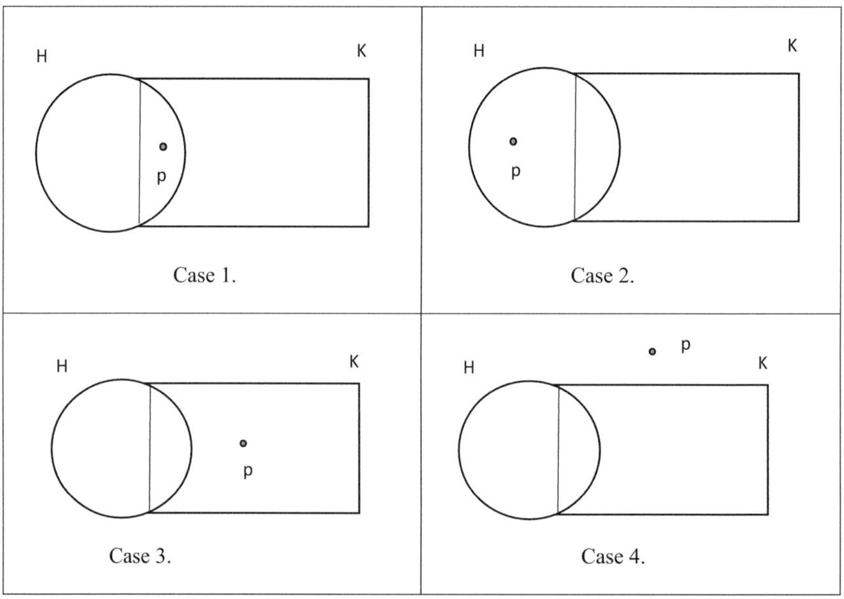

Fig. 6.2 Set theory approach for disjunction

Table 6.5 Truth table of conditional statement

h	m	$h \to m$
T	T	T
T	F	F
F	T	T
F	F	T

Note The two logical statements can give the meaningful interpretations. Otherwise, the conditional statement will become the irrelevant or meaningless. For example, $h =$ Atharva is a clever student and $m = K$ is the capital of State. The conditional statement of h and m is as follows

6.4.4 Conditional Statement (If… then…)

A conditional statement of two simple statements h and m is denoted by $h \to m$ and it is read as "If h then m" or "h only if m" or "h implies m." In the statement $h \to m$, the statement h is called the Hypothesis or antecedent and m is called conclusion or consequent [1].

The truth value of $h \to m$ is false if h is true and m is false, Otherwise it is true.

The truth table of the conditional statement is depicted in Table 6.5

"If Atharva is a clever student, then K is the capital of India." There is no any relation between h and m. That is, h and m are not logical statements.

Examples Example 6.6: Let $H \subset K \subset U$. The following diagram depicts the all possible cases of the set and it's subset.

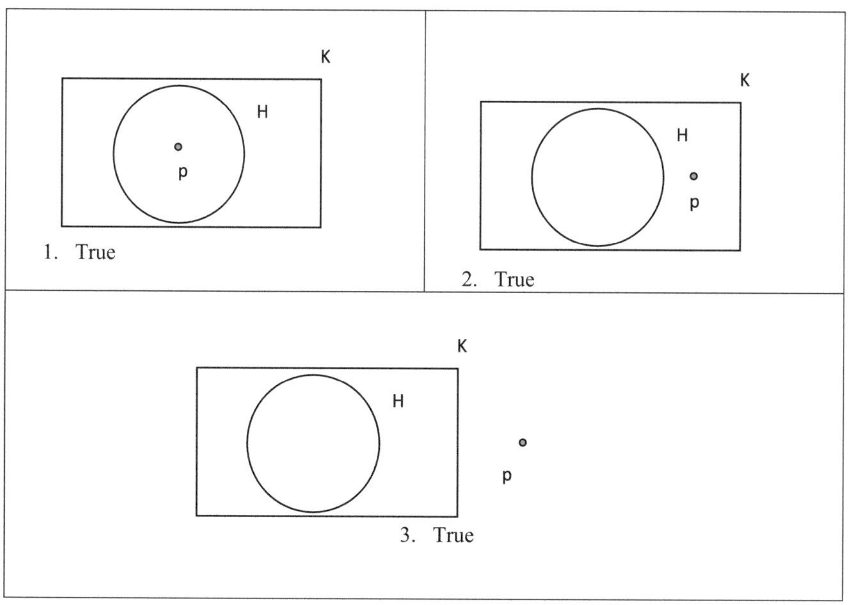

Fig. 6.3 Set theory approach for conditional statements

Table 6.6 Truth table of conditional statement

h	m	$h \leftrightarrow m$
T	T	T
T	F	F
F	T	F
F	F	T

6.4.5 Biconditional Statement (If and only if or iff)

A Biconditional statement of two simple statements h and m is denoted by $h \leftrightarrow m$ and it is read as "h if and only if m" or "h iff m" or "h implies and implied by m" or "h is necessary and sufficient for m." The truth value of $h \leftrightarrow m$ is true if h and m have the same truth value, i.e., either both are true or false. Otherwise, it is false [1].

The truth table of the biconditional statement is depicted in Table 6.6.

Examples

1. Let h: $x = 5$ and m: $x + 13 = 18$.

We have, if $x = 5$ then $x + 13 = 5 + 13 = 18$ and if $x + 13 = 18$ then $x = 5$. Thus, $x = 5$ iff $x + 13 = 18$.

2. Find the truth table for "An integer is an even iff it is divisible by 2."

Let h: v is an integer divisible by 2 and m: v is an even number.

Consider the following table.

Sr. No.	h	m	Biconditional statement	$h \leftrightarrow m$
1.	122 is an integer divisible by 2 (**True**)	122 is an even number (**True**)	122 is an integer divisible by 2 iff 122 is an even number (**True**)	**T**
2.	12 is an integer divisible by 2 (**True**)	12 is an even number (**False**)	12 is an integer divisible by 2 iff 12 is not an even number (**False**)	**F**
3.	200 is an integer divisible by 2 (**False**)	200 is an even number (**True**)	200 is not divisible by 2 iff 200 is an even number (**False**)	**F**

(continued)

(continued)

Sr. No.	h	m	Biconditional statement	$h \leftrightarrow m$
4.	26 is an integer divisible by 2 (**False**)	26 is an even number (**False**)	26 is an integer not divisible by 2 iff 26 is not an even number (**True**)	**T**

6.4.6 *Special Statements*

If $h \rightarrow m$ is a conditional statement then

1. It's converse statement is $m \rightarrow h$.
2. It's inverse statement is ~ h → ~ m.
3. It's contrapositive statement is ~ m → ~ h.

6.5 Tautology and Contradiction: Experiential Learning

There are numerous circumstances in which either all statements are correct or all statements are false. Suppose in university examinations a particular student gets distinction in all subjects or a particular student don't get distinction in any subject.

Such types of similar situations are represented by tautology and contradiction in Mathematics.

Tautology: A statement formula which is always true, for all possible values of its propositional variables is called a tautology.

Let's consider Table 6.7.

Therefore, the statement formulas h ∨ ~ h, $h \rightarrow h$ is tautology.

Contradiction: A statement formula is false, for all possible values of its propositional variables is called a contradiction.

A formula h ∧ ~ h is a contradiction.

Contingency: A statement formula that is neither tautology nor contradiction is called a Contingency.

Table 6.7 Connectives in mathematics

h	$\sim h$	$h \vee \sim h$	$h \rightarrow h$	$\sim (h \wedge \sim h)$
T	F	T	T	T
F	T	T	T	T

6.6 Logical Equivalence

In real life, we have observed that there are numerous similar things with respect to particular attributes. That is, two cars are similar with respect to their average. Two students are similar with respect to their height and weight. Likewise, in logic, we can say that two formulas are similar with respect to their truth values.

Two statements are logically equivalent iff they have the same truth values, for all possible choices of the truth values of the statements involved in it. Two formulas are equivalent even if they have different variables [1, 6].

If P is equivalent to Q then it is represented as $P \Leftrightarrow Q$ or $P \equiv Q$.

The symbol $\Leftrightarrow$ is not connective. So, $\Leftrightarrow$ and $\leftrightarrow$ have different meanings in the logic theory.

For example, $h \to (m \vee p) \equiv (h \to m) \vee (h \to p)$. The readers can prove this by truth table.

Logical Identities: The following are some logical identities.

Sr. No.	Name of identity	Expression of identity
1.	Idempotence of $\vee$	$p \equiv p \vee p$
2.	Idempotence of $\wedge$	$p \equiv p \wedge p$
3.	Commutativity of $\vee$	$h \vee m \equiv m \vee h$
4.	Commutativity of $\wedge$	$h \wedge m \equiv m \wedge h$
5	Associativity of $\vee$	$p \vee (h \vee m) \equiv (p \vee h) \vee m$
6.	Associativity of $\wedge$	$p \wedge (h \wedge m) \equiv (p \wedge h) \wedge m$
7.	Distributivity of $\wedge$ over $\vee$	$p \wedge (h \vee m) \equiv (p \wedge h) \vee (p \wedge m)$
8.	Distributivity of $\vee$ over $\wedge$	$p \vee (h \wedge m) \equiv (p \vee h) \wedge (p \vee m)$
9.	Double Negation	$h \equiv \sim (\sim h)$
10.	De Morgan's Laws	$\sim (h \vee m) \equiv \sim h \wedge \sim m$
11.	De Morgan's Laws	$\sim (h \wedge m) \equiv \sim h \vee \sim m$
12.	Tautology	$h \vee \sim h \equiv T$
13.	Contradiction	$m \wedge \sim m \equiv F$
14.	Absorption Laws	$h \vee (h \wedge m) \equiv h$
15.	Absorption Laws	$h \wedge (h \vee m) \equiv h$

6.7 Algebra of Propositions

The conjunctions (product) of the variables and their negations is called an elementary conjunction or fundamental conjunctions. If p and q are statements then h, ~ h, m, ~ m, $h \wedge m$, $\sim h \wedge m$, $h \wedge \sim m$, $\sim h \wedge \sim m$ are called elementary conjunctions.

The disjunctions (or sum) of variables and their negations is called an elementary disjunction or fundamental disjunctions. The expressions or formulae h, ~ h, m, ~ m, $h \vee m$, $\sim h \vee m$, $h \vee \sim m$, $\sim h \vee \sim m$ are called elementary disjunctions.

A structured representations of logical expressions according to some rule is called a **Normal form**.

Generally, there are two types of normal form depending upon the nature of conjunctions and disjunctions.

6.7.1 *Disjunctive Normal Form (DNF)*

An expression or formula, which involves of a disjunction of fundamental conjunctions or elementary products. It is abbreviated as DNF.

The expressions $(h \wedge m) \vee m \vee (h \wedge \sim m)$ and $(h \wedge r) \vee (q \wedge h) \vee (\sim m)$ are examples of Disjunctive normal forms [1].

A disjunctive normal form of any formula is obtained by using the following ways.

1. In the formula, replace $\rightarrow$ *and* $\leftrightarrow$ by the by expressions using the logical connectives $\wedge$, $\vee$, $\sim$. We have, $h \rightarrow n \equiv \sim h \vee m$ and $h \leftrightarrow m \equiv (\sim h \vee m) \wedge (\sim m \vee h)$.
2. Use formula to eliminate " ~ " before sums or products. That is, $\sim (h \vee m) \equiv \sim h \wedge \sim m$ and $\sim (h \wedge m) \equiv \sim h \vee \sim m$.
3. Apply distributive laws continually.

6.7.2 *Conjunctive Normal Form (CNF)*

An expression or formula that involves of a conjunction of fundamental disjunctions or elementary sums. It is denoted as CNF.

The expressions $(h \vee m) \wedge (h \vee \sim m)$ and $(\sim h \vee r) \wedge (q \vee h) \wedge (\sim r \vee h)$ are examples of conjunctive normal forms.

6.8 Conditional and Biconditional Statements: Experiential Learning

(1) **Conditional Statement: Profound Insights**

Let us identify the situations having the first statement h is the necessary condition for the second statement m. Moreover, h is the only necessary condition for m but not sufficient condition for existence of the statement m. We can write this as $h \rightarrow m$.

For example, let h: x is an integer having zero at it's unit place and m: x is an even number. Here, h is the necessary but not sufficient condition for m. If x is an integer having zero at it's unit place, then x is an even number. However, the integer 14 has not zero digit at it's unit place, but 14 is an even integer. We represent this as $h \rightarrow m$.

What will happen if we take m as the sufficient condition. For example, h: **x** is divisible by 2 and m: x is an even integer. In the conditional statement, if h is F and m is T then $h \rightarrow m$ is T. But, in this example, h is false and m is true means x is not divisible by 2 and x is an even integer. Furthermore, $h \rightarrow m$ for this case is "if x is not divisible by 2 then x is an even number." This statement is false. The actual meaning of $h \rightarrow m$ is false but by the truth table it is true. Hence, h must be only necessary condition for the existence of m.

(2) Bic**onditional Statement: Profound Insights**

The statement h is the necessary and sufficient condition for the existence of the statement m. That is, $h \rightarrow m$ and $m \rightarrow h$ iff $h \leftrightarrow m$.

Moreover, $h \leftrightarrow m \equiv (h \rightarrow m) \wedge (m \rightarrow h)$.

Suppose h: an integer x is a multiple of 2 and m: an integer x is an even. Here, h is the necessary and sufficient condition for the existence of m. So, if any one of them is false then other is also false.

In the biconditional statements, h must be the necessary and sufficient condition for the existence of statement m.

6.9 Arguments: Experiential Learning

(A) **Brainstorming Experiment-III**

Let us identify the circumstances or situation in which all final conclusions are true for all possible values of variables. For example, consider the statements, h: w is a multiple of 4, m: w is a multiple of 2 and p: w is an even integer.

(B) **Inferences-III**

1. Suppose h is true and $h \rightarrow m$ is also true. Therefore, w is a multiple of 4 and if w is a multiple of 4 then w is a multiple of 2. This implies w is a multiple of 2. Thus, h and $h \rightarrow m$ are true then these together implies that m is true. **(Law of Detachment)**
2. Suppose $h \rightarrow m$ is true and ~ m is true. So, "if w is a multiple 4 then w is a multiple of 2" is true and w is not multiple of 2 is true then w is not multiple of 4. Therefore, $h \rightarrow m$ and ~ m together implies ~ h. That is, $(h \rightarrow m) \wedge \sim m \rightarrow \sim h$ is a tautology. (**Rule of Contradiction)**
3. Suppose $h \rightarrow m$ and $m \rightarrow p$. Therefore, if w is a multiple 4 then w is a multiple of 2 and if w is a multiple of 2 then w is an even number. This implies that, if w is a multiple 4 then w is an even number, i.e., $h \rightarrow p$. This rule is known as Hypothetical Syllogism

(C) Mathematical Insights-III

Arguments

In the previous sections, we have discussed about the propositions, truth table and other aspects. Here and now, we will shade some light on the methods that help us to connect or link the statements to design a logically valid argument.

A valid argument is a finite sequence of statements $h_1, h_2, h_3, \cdots, h_n$ called as premises or assumptions or hypothesis together with a statement C, called as the conclusion such that $(h_1 \wedge h_2 \wedge h_3 \wedge \cdots \wedge h_n) \rightarrow C$ is a tautology.

There are several methods for validating the theorem based on the rules of inferences, that are given below [1, 3].

4. **Law of Detachment (Modus Ponens)**

Whenever, the statements h and $h \rightarrow m$ are true, then the statement m is also true. This rule can be represented as follows.

$$\begin{array}{c} h \\ h \rightarrow m \\ \hline \therefore\ m \end{array}$$

The assertions above the horizontal line represents premises or hypothesis and the below line is called as the conclusion.

This rule creates the valid argument as " $h \wedge (h \rightarrow m) \rightarrow m$ is a tautology."

As the conclusion m is detached from the premise h and $h \rightarrow m$, so it is called as the law of detachment or Modus Ponens.

5. **Law of Contrapositive (Modus Tollen)**

Whenever $h \rightarrow m$ is true and m is false then h is false, is known as the law of Contrapositive or Modus Tollen. This is represented as follows.

$$\begin{array}{c} h \rightarrow m \\ \sim m \\ \hline \therefore\ \sim h \end{array}$$

This rule constitutes the valid argument as "$(h \rightarrow m) \wedge \sim m \rightarrow \sim h$ is a tautology".

6. **Disjunctive Syllogism**

This rule states that "*if h*$\vee$*mis true and h is false then m must be true*" It is represented in the following form as

$$\begin{array}{c} h \vee m \\ \sim h \\ \hline \therefore\ m \end{array}$$

This rule constitutes the valid argument as "$(h \vee m) \wedge \sim h \rightarrow\ m$ is a tautology".

7. **Hypothetical Syllogism**

This rule states that "*if* $h \rightarrow$ *mand* $m \rightarrow$ *pare true and* $h \rightarrow$ *pis true*" It is represented in the following form as

$$\frac{\begin{array}{c} h \rightarrow m \\ m \rightarrow p \end{array}}{h \rightarrow p}$$

This rule institutes the valid argument as "$(h \rightarrow m) \wedge (m \rightarrow p) \rightarrow\ (h \rightarrow p)$ is a tautology".

Example 6.7 Determine the validity of the statement

x: All my friends are musicians
y: John is my friend
z: None of my neighbours are musicians
---
c: John is not my neighbour

Solution Let,
h: All my friends are musicians
m: John is my friend
p: None of my neighbors are musicians
s: John is my neighbor
Therefore,

$$\frac{\begin{array}{c} h \\ m \\ \sim p \end{array}}{\therefore\ \sim s}$$

As all my friends are musicians and John is my friend $\Rightarrow$ John is a musician.
$\therefore\ h \wedge m\ \Rightarrow$ John is musician
$h \wedge m \wedge \sim p\ \Rightarrow$ John is musician and my neighbors are not musicians
$\therefore h \wedge m \wedge \sim p\ \Rightarrow$ John is not my neighbor
$\therefore h \wedge m \wedge \sim p\ \rightarrow \sim s$ is true
Therefore, the given statement is valid.

6.10 Propositional Functions, Quantifiers: Experiential Learning

(A) Brainstorming Experiment-IV

Mathematics is a universally accepted language. Every mathematical symbol has a unique meaning across the globe. Thus, Mathematics is the only language used in every part of the world. Ask the students to write all mathematical symbols and their meanings. Identify the mathematical symbols that are useful in the discrete mathematics.

(B) Inferences-IV

Two phrases are very important in the discrete mathematics.

i. For all: It is mathematically denoted by $\forall$. (Universal Quantifier)
ii. There exists: It is mathematically denoted by $\exists$. (Existential Quantifier)

(C) Mathematical Insights-IV

The predicate h involving k variables $h_1, h_2, h_3, \ldots, h_k$ is called k-place predicate and it is denoted by $h(h_1, h_2, h_3, \ldots, h_k)$, where each variable h_i is called as argument.

The word quantifier refers to an amount of quantity in general. But, in logic it has only two meanings "for all and for some." [1]

Types of Quantifiers

i. **Universal Quantifier**

If $h(x)$ is a predicate and x is an argument, then the universal quantifier is stated as follows.

"For all values of x, $h(x)$ is true." The phrase "For all" is denoted by $\forall$.

$\forall$ means for every or for each or for all. Notation:

"$\forall x\ h(x)$ is true".

ii. **Existential Quantifier**

In real life situations, the predicate is right for only some values of argument. Suppose $\forall x\ h(x)$ is false, but there exists x for which $h(x)$ is true. Then, we say that x is bounded by existential quantification.

The phrase "there exists x" is called an existential quantifier. It is denoted by $\exists$.

$\exists x\ p(x)$ is true means there exists x, such that $p(x)$ is true or for at least one value of x, $p(x)$ is true.

Note In this context, at least one x means there are some values of x but not all values.

6.11 Negation of Quantified Statements: Experiential Learning

(A) **Brainstorming Experiment-V**

Suppose, a teacher is teaching a subject of discrete mathematics to students of the second year B. Tech. program. The teacher makes a statement that "All people present in the class are students of the second year B.Tech. program." He asked the students, whether this statement is true or false. A student told that this statement is not true as you are present in the class but you are not a student of the B. Tech. program. That is "not all people present in the class are students of the second year B.Tech. program." This means that there is at least one who is not a student of this program.

(B) **Inferences-V**

i. The negation of "all" is "not all" or "there exists at least one for which the result is not true."
ii. The negation of "there exists" is "for all"

(C) **Mathematical Insights-V**

Consider the statement $\forall x\ h(x)$. Its negation is "It is not the case that for all x, $h(x)$ is true." That means for some $x = a$, $h(a)$ is not true OR $\exists\, x, \sim h(x)$ is true. The following table depicts the statements and it's negations [1].

Statement	It's negation
$\forall w\ [h(w)]$	$\exists\, w,\ [\sim h(w)]$
$\exists\, w,\ [\sim h(w)]$	$\forall w\ [h(w)]$
$\forall w\ [\sim h(w)]$	$\exists\, w\ [h(w)]$
$\exists\, w\ [h(w)]$	$\forall w\ [\sim h(w)]$

Rules of Inferences

1. Rule I: Universal Instantiation

$$\frac{\forall w\ h(w)}{\therefore\ h(k)}$$

where, k belongs to the universal set.

2. Rule II: Existential Instantiation

$$\frac{\exists w\ h(w)}{\therefore\ h(k)}$$

where, k is in the universal set for which $h(k)$ is true.

3. Rule III: Universal Generalization

$$\frac{h(w)}{\forall\ w\ h(w)}$$

4. Rule IV: Existential Generalization

$$\frac{h(a)}{\exists\ wh(w)}$$

where k is in the universal set.

(I) **Quantifiers in two variables**

Quantifiers	Expressions
$\exists u\ \forall w\ [h(u, w)]$	$\exists$ a value of u such that for all values of w, $h(u, w)$ is true
$\forall w\ \exists u\ [h(u, w)]$	For each value of w, $\exists$ a value of u such that $h(u, w)$ is true
$\exists u\ \exists w\ [h(u, w)]$	$\exists$ a value of u and value of w such that $h(u, w)$ is true
$\forall u\ \forall w\ [h(u, w)]$	For all values of u and w, $h(u, w)$ is true

Examples Example 6.8: Let h (w): w is an even (Multiple of 2)
$m(w)$: w is a prime
$r(u, w)$: $u + w$ is an even integer
Using this information rewrite, the following sentences in the symbolic form.

(i) Every integer is an odd integer
(ii) Every integer is either even or prime
(iii) Addition of any two integers is an odd integer.

Solution

(i) $\forall\ w\ [\sim\ h(w)]$
(ii) $\forall\ w\ [h(w) \vee m(w)]$
(iii) $\forall\ u \forall w\ [\sim\ r(u, w)]$.

6.12 Case Study and Projects

Consider the following case study/projects for deepen the understanding of the concepts.

1. Write the list of all Mathematical symbols, their meanings, and when to use the concerned symbol.
2. How are the concepts logic useful in designing the digital electronic circuit?

3. Develop a game that involves the concepts of logic.
4. Design an experiment by using the concepts of logic and explain how this theory helps in making the precise inferences and boosting decision-makings.
5. Importance of basic logical operations for designing the railway routes.

Exercise

1. Define the following and explain at least one example of each.

    a. Proposition
    b. Conditional statement
    c. Biconditional statement
    d. Conjunction statement
    e. Disjunction statement
    f. Tautology
    g. Contradiction
    h. Negation of quantifiers.

2. Prove all logical identities given in the Sect. 6.6.
3. Rewrite the following statements by using the quantifier variables and predicate symbols

    a. All animals can fly
    b. No hen can fly
    c. Some animals can fly
    d. Not all animals can fly
    e. There are animals that cannot fly.

4. Explain the concepts of Conditional and biconditional statements by using the sets and subsets.
5. Explain the concepts of DNF and CNF and design your own examples of these concepts.

References

1. Peter J (2000) Sets, logic and categories. Springer. ISBN-9781852330569
2. Aigner M (2023) Discrete mathematics. American Mathematical Society
3. Erciyes K (2021) Discrete mathematics and graph theory. Springer, Cham. https://doi.org/10.1007/978-3-030-61115-6
4. Rosen K, Krithivasan K (2021) Discrete mathematics and its applications. Mc Graw Hill, India. ISBN: 978930727353
5. Jiri M, Jaroslav N (2011) Invitation to discrete mathematics, 2nd edn. Oxford University Press. ISBN: 9780198570424
6. Tremblay JP, Manohar R (2023) Discrete mathematical structures with applications to computer science. McGraw Hill Education. ISBN: 9780074631133

Chapter 7
Permutations and Combinations

Mathematics is not about numbers, equations, permutations, computations, or algorithms: it is about understanding.
—William Paul Thurston

Abstract Mathematics is not about numbers, equations, permutations, computations, or algorithms: it is about understanding. There was a special arrangement of soldiers used in the greatest war of the Mahabharata in India. People who didn't know mathematical theory have been using the concepts of permutations and combinations to resolve their day-to-day issues in more pragmatic ways. This chapter deals with the concept of basic counting principles by using the brainstorming experiment of the selection of class representatives from the class. The concept of permutation and related results are explained by using the concepts selection of Mathematicians for four prestigious awards namely as per ranking, Abel Prize, Field Medal, Chern Medal and Wolf Prize at international level. Different real-life examples demonstrate the concept of the combination and concerned types. Moreover, special applications highlight the importance of the concepts and enhance the understanding of the students. The chapter accomplishes with case studies and exercises designed to bridge the gap between theoretical relational structures and their multidisciplinary applications.

Keywords Counting principles · Basic mathematical functions · Permutations · Combinations

7.1 Introduction

Numerous circumstances that involve the arrangement of objects or the combination of objects. In ancient times, the combinations of different animals and soldiers were used during the war. By considering the actual experiences, the various combinations are used for getting the success in that activity. There was a special arrangement of soldiers used in the greatest war of the Mahabharata in India. People who didn't know

© The Author(s), under exclusive license to Springer Nature Singapore Pte Ltd. 2026

H. R. Bhapkar and P. N. Mahalle, *Discrete Mathematics: An Experiential Learning Perspective*, Studies in Autonomic, Data-driven and Industrial Computing,
https://doi.org/10.1007/978-981-95-7210-6_7

the mathematical theory have been using the logic of permutations and combinations to resolve their day-to-day issues in more pragmatic ways.

In this chapter, we deal with the concepts of permutations, combinations, their different cases, and related examples [1–3].

7.2 Basic Counting Principles: Experiential Learning

Let's consider the following brainstorming Experiment-I for understanding the concept of the basic counting principles.

(A) Brainstorming Experiment-I

Suppose there are 64 students in a class of the first year of a particular program. Out of these students, there are 40 boys and 24 girls. The head of the department wants to select two students' representatives from the class. There are two ways to select the class representatives.

1. Select any two students from the class.
2. Select one boy and one girl students from the class.

In how many ways the concerned authorities can select the class representatives. Let's consider the following inferences or observations.

(B) Inferences-I

In Experiment-I, we have observed the following special inferences.

1. In the first case, there are 64 possible alternatives for selecting the first class representative and 63 alternatives for the second-class representative. Therefore, the total number of ways for selecting two class representatives is $64 * 63 = 4032$. This philosophy leads to the concept of the "PRODUCT RULE" of the counting principle.
2. There are 40 choices for selecting a boy representative of the class and 24 choices for selecting a girl representative in the case 2. The total number of ways are $40 + 24 = 64$.

This concept demonstrates the "SUM RULE" of the counting principle.

(C) Mathematical Insights-I

In Mathematics, there are two basic operations, Multiplication and Addition. By the flavors of these operations, there are two basic counting principles, known as the product rule and sum rule.

The Product Rule (The Principle of Sequential Counting)

Suppose, a process can be divided into a sequence of two jobs. If the first job can be performed in h_1 ways and the second job can be performed in h_2 ways then there are $h_1 h_2$ ways to complete the process.

We have, C and H are two non-empty sets with $|C| = m$ and $|H| = n$ then $|C \times H| = m \times n$, where $C \times H$ is the Cartesian product of C with H [4].

Examples

Example 7.1 *In Experiment-I, the inference (1) is an example of the product rule of the counting principle.*

Example 7.2 *How many different number plates of the vehicle are there, if each plate involves a sequence of four alphabets followed by three digits? (The number must be non-zero).*

Solution There are 26 alphabets and 10 digits. Therefore, the first four alphabets are selected by $26 \times 26 \times 26 \times 26$ ways and the next three digits are selected by $10 \times 10 \times 9$ ways. Hence, the required number of ways are $26 \times 26 \times 26 \times 26 \times 10 \times 10 \times 9 = 411278400$. For example

A	*A*	*A*	*A*	**0**	**0**	**1**

Extended Product Rule

Suppose the procedure is divided into a sequence of n tasks as $T_1, T_2, T_3, \cdots, T_m$. If each task T_i can be done in n_i ways, for $i = 1, 2, 3, \cdots, m$, then there are $n_1 \cdot n_2 \cdot n_3 \cdots n_m$ ways to complete the procedure. Therefore, $|T_1 \times T_2 \times T_3 \times \cdots \times T_m| = n_1 \times n_2 \times n_3 \times \cdots \times n_m$.

Example 7.3 *If $f : C \to H$ is a function with $|C| = m$ and $|H| = n$ then prove that there are n^m functions.*

Solution: There are n choices in codomain set H for every element of the domain set C. Hence, by the product rule, there are $n \times n \times n \times \cdots m$ times $= n^m$ function from the C to H.

Example 7.4 *How many one–one functions are there if $f : C \to H$ is a function with $|C| = m$ and $|H| = n$.*

Solution: Case1: If $m > n$ then there is no one–one function from C to H.

Case 2: Let $m \leq n$. Let $C = \{c_1, c_2, c_3, \ldots, c_m\}$. There are n ways to choose an image of c_1. As the function is one–one, so there are $n - 1$ ways to choose an image of c_2 and $n - 2$ ways for the image of c_3. Similarly, there are $n - (r - 1)$ ways for the image of c_r. This is the sequential procedure for selecting the images of all elements of the domain set. Therefore, there are $n\,(n - 1)\;(n - 3)\;(n - 4)\;\cdots (n - m + 1)$ 1–1 functions from C to H.

Example 7.5 *In how many ways a student can answer all questions in an examination involving 13 true or false questions.*

Solution: There are two ways of answering each question $\{T, F\}$. By the product rule, the number of ways in which all questions are answered is $2 \times 2 \times 2 \times 2 \times 2 \times 2 \times 2 \times 2 \times 2 \times 2 \times 2 \times 2 \times 2 = 2^{13}$.

The Sum Rule

Suppose that a procedure is divided into two disjoint tasks. If the first task can be executed in h_1 ways and the second task can be executed in h_2 ways then there are $h_1 + h_2$ ways to complete the procedure.

Analogically, by using the set theory, if there are two disjoint sets C and H such that $|C| = m$ and $|H| = n$ then $|C \cup H| = m + n$ [5].

Example 7.6 *In Experiment-I, the inference (2) is an example of the sum rule of the counting principle.*

Example 7.7 *How many ways can we get a sum of 5 or 11 when two different dice are rolled?*

Solution Two dice are different, therefore, (p, q) and (q, p) are different when $p \neq q$.

Let E_1 be the set of ordered pairs having sum 5.
$\therefore \quad E_1 = \{(1, 4), (2, 3), (3, 2), (4, 1)\}$. There are 4 ways to get sum 5.
Let E_2 be the set of ordered pairs having sum 11.
$\therefore \quad E_2 = \{(5, 6), (6, 5)\}$. There are 2 ways to get the sum 11.
The sets E_1 and E_1 are mutually exclusive, i.e., $E_1 \cap E_2 = \varphi$.
Therefore, by the sum rule, the number of ways to get sum 5 or 11 is $4 + 2 = 6$.

Example 7.8 *How many possible passwords are there for a laptop if password involves 6 to 8 characters long in which each character is capital alphabet or digit and each password has at least one digit?*

Solution Let P be the total number of possible password with given conditions. Let P_1, P_2, and P_3 be the possible passwords of length 6, 7, and 8, respectively. Therefore,

$$P = P_1 + P_2 + P_3 \ldots (1)$$

There are 26 alphabets and 10 digits. Therefore, P_1 is calculated by finding the number of passwords of length 6 with alphabets and digits and exclude the number of passwords having alphabets only. Therefore,

$$P_1 = (26 + 10)^6 - 26^6 = 2176782336 - 308915776 = 1867866560.$$

Similarly,

$$P_2 = (26 + 10)^7 - 26^7 = 78364164096 - 8031810176 = 70332353920$$

$$P_3 = (26 + 10)^8 - 26^8 = 2821109907456 - 208827064576 = 2612282842880$$

$$P = P_1 + P_2 + P_3 = 2684483063360$$

7.3 Mathematical Functions: Experiential Learning

As every theory needs some basic foundations, the permutations and combinations require the concept of factorial functions. The factorial of a natural number is the foundation of the counting theory.

Factorial

Let n be any natural number. The factorial of n is defined as,

$$n\,! = n\,(n-1)\,(n-2)\,(n-3)\cdots(3)\,(2)(1)$$

For example:

$$2! = 2 \times 1 = 2$$
$$3! = 3 \times 2 \times 1 = 6$$
$$4! = 4 \times 3 \times 2 \times 1 = 24$$
$$5! = 5 \times 4 \times 3 \times 2 \times 1 = 120$$

And so on.

Notes

1. Zero factorial is 1, i.e., $0\,! = 1$
2. $p\,! = p \times (p-1)! = p \times (p-1) \times (p-2)!$

$$100\,! = 100 \times 99\,! = 100 \times 99 \times 98 \times 97 \times 96\,!$$

3. $\frac{p\,!}{(p-3)\,!} = p \times (p-1) \times (p-2)$

7.4 Permutations: Experiential Learning

(A) Brainstorming Experiment-II

In Mathematics, there are mainly four prestigious awards namely as per ranking, Abel Prize, Field Medal, Chern Medal, and Wolf Prize at international level. Suppose, the committee wants to offer one award to one Mathematician out of 12 top Mathematicians at international level. The research work of these 12 Mathematicians excellent and incomparable. How many ways the committee can choose the Mathematicians.

(B) Inferences-II

We have to select 4 out of 12 Mathematicians. The first Mathematician can be nominated in 12 ways. The second, third, and fourth Mathematicians can be nominated in 11, 10, and 9 ways, respectively.

Therefore, the number of ways to select awardee is $12 \times 11 \times 10 \times 9 = 11880$. This philosophy leads to the concept of the "PERMUTATIONS"

(C) Mathematical Insights-II

In this chapter, we use mainly two words "Arrangements" and "Selections." In arrangements, the order of the objects is important, it is not a set whereas in the selection, the order of the objects in immaterial, it is a set. Arrangement is associated with the Permutation and the selection with combinations.

Definition An ordered arrangement of objects/elements of a set of distinct objects/ elements is called a permutation.

Depending upon the nature of arrangements, there are three types of the Permutations.

Type I: Permutations when all elements are distinct

A permutation of h distinct entities taken m at a time is an arrangement of m objects out of h objects, where $h \geq m$. It is denoted by $P(h, m)$ *or* ${}^hP_m$ [4].

Theorem 7.1 *The number of m-permutations on h distinct objects is*

$$P(h, m) \text{ or } {}^hP_m = h \times (h-1) \times (h-2) \times (h-3) \times \cdots \times (h-m+1) = \frac{h!}{(h-m)!}$$

Proof Let h be the number of distinct objects in a set say $O_1, O_2, O_3, \ldots, O_h$. The first element of a permutation can be chosen in h ways. After filling the first position, the second element can be nominated in $h-1$ ways, the third in $h-2$ ways and so on. The m-th position element can be selected in $[h-(m-1)] = h-m+1$ ways.

Hence, by the product rule,

$$P(h, m) = h \times (h-1) \times (h-2) \times \cdots \times (h-m+1)$$
$$P(h, m) = \frac{h \times (h-1) \times (h-2) \times \cdots \times (h-m+1) \times [(h-m)!]}{(h-m)!} = \frac{h!}{(h-m)!}$$

Notes:

i. $P(h, h) = \frac{h!}{(h-h)!} = \frac{h!}{(0)!} = h! \quad (\because o! = 1)$

ii. $P(k, 1) = \frac{k!}{(k-1)!} = k$
iii. $P(u, 2) = \frac{u!}{(u-2)!} = u \times (u-1)$
iv. $P(w, 3) = \frac{w!}{(w-3)!} = w \times (w-1) \times (w-2)$

Examples

Example 7.9 Find $P\,(h, r)$, $h = 5$ and

(i) $r = 2$ (ii) $r = 3$ (iii) $r = 4$ (iv) $r = 5$

Solution: Here $h = 5$.

(i) $P\,(5, 2) = {}^5P_2 = \frac{5!}{(5-2)!} = 20.$
(ii) ${}^5P_3 = \frac{5!}{(5-3)!} = 60$
(iii) ${}^5P_4 = \frac{5!}{(5-4)!} = 120.$
(iv) ${}^5P_5 = \frac{5!}{(5-5)!} = 120.$

Example 7.10

i. *How many numbers of the form xyzw can be made with numbers 1, 2, 3, 5, 7, 8 if digit repetitions are not allowed?*
ii. *How many such numbers are < 5000?*
iii. *How many in (i) are even?*
iv. *How many in (ii) are odd?*
v. *How many in (i) are contain both 3 and 5?*
vi. *How many in (i) are divisible by 10?*

Solution i. There are six distinct numbers and out of 6 numbers, 4 digit numbers can be formed in ${}^6P_4 = \frac{6!}{(6-4)!} = 360.$

ii. The four digit numbers which are less than 5000 are the numbers in which the first digit 1 or 2 or 3. Therefore, the first digit can be chosen in 3 ways, the second digit can be chosen in 5 ways, the third and fourth digits can be chosen 4 and 3 ways, respectively. Hence, the required number of 4 digit numbers are $3 \times 5 \times 4 \times 3 = 180$.
iii. The numbers in (i) are even if their last digits are 2 or 8. Hence, the total required numbers are $2 \times 5 \times 4 \times 3 = 120$.

Place of digit	Thousandth	Hundredth	Tenth	Unit
Number of ways	3	4	5	2

iv. Numbers less than 5000 and odd are obtained by selecting the odd digit at the unit place in (ii). The odd numbers are 1, 3, 5, 7.

Place of digit	Thousandth	Hundredth	Tenth	Unit
Possible digits	{1, 2, 3}	{1, 2, 3, 5, 7, 8}	{1, 2, 3, 5, 7, 8}	{1, 3, 5, 7}

Therefore, as digit repetitions are not allowed, there are four possible cases.

Digit at unit place	Number of possible ways				
	Unit	Hundredth {1,2,3}	Tenth	Thousandth	Total numbers
1	1	2	4	3	24
3	1	2	4	3	24
5	1	3	4	3	36
7	1	3	4	3	36

Hence, the required numbers are $24 + 24 + 36 + 36 = 120$.

v. In the four-digit number, any two positions are occupied by 3 and 5 and the remaining two positions are occupied by 1, 2, 7, 8. The number of ways in which two positions are occupied by 3 and 5 will be $4 \times 3 = 12$. Now, the remaining two positions will be occupied by the remaining digits, i.e., 1, 2, 7, and 8. Out of two positions, the first position is occupied by 4 ways and the second by 3 different ways. Hence, the total numbers are $12 \times 4 \times 3 = 144$.
vi. We know that the number is divisible by 10 iff zero is present at it's unit place. But, there is no zero digit in the given set of numbers. Hence, there is no any such number.

Example 7.11 *There are 10 different books of DM, 5 different books of DS, 3 different books of EM, and 7 different books of LA. How many different arrangements are possible if*

i. *The books of the same course must be together.*
ii. *Only DM books must be together.*

Solution The total number of books are 25.

i. The books of DM can be organized among themselves in 10! ways, DS books in 5! ways, EM books in 3! Ways, and LA books in 7! Ways. Hence, the total number of arrangements are $10! \times 5! \times 3! \times 7!$.
ii. Consider the 10 DM books as a single book. So, there are 16 positions to make arrangements of books. Therefore, the number of arrangements are 16!. In these arrangements, the DM books can be arranged themselves in 10! ways. Hence, the total number of arrangements are $10! \times 16!$

Example 7.12 *There are 2 major papers of Mathematics, 2 papers of minor, 1 paper of SEC, 2 papers of MDC in the first semester of B.Sc. Honors program, are to be arranged at an examination. Find the total number of ways if*

1. *Mathematics papers are consecutive.*
2. *MDC papers are not consecutive.*

Solution The total number of papers are 7.

1. Both Mathematics papers must be consecutive. So, consider two papers as a single paper. These two Mathematics papers can be organized between themselves in 2! ways. Therefore, we need to arrange 6 papers in different ways. These 6 papers can be decided in 6! different ways. Hence, the total number of arrangements $= 2! \times 6!$ ways.
2. There are 2 MDC papers and 5 other papers (Non-MDC). Two MDC papers M_1 and M_2 are to be arranged among 4 gaps or at two ends as depicted in the following figure.

MDC M_1 or M_2	Other Paper	MDC M_1 or M_2	Other Paper	MDC M_1 or M_2	Other Paper	MDC M_1 or M_2

Therefore, there are 6 places, where MDC papers can be arranged. So, the MDC papers can be arranged in ${}^6P_2 = 30$ ways. The remaining five papers can be arranged in 5! ways. Hence the total number of arrangements are $= {}^6P_2 \times 5! = 30 \times 120 = 3600$.

Example 7.13 *Find the value of h if.*

1. $P(h, 2) = 20$
2. $2P(h, 2) + 98 = P(2h, 2)$

Solution

1. We have

$$\begin{aligned} &P(h, 2) = 20 \\ &\Rightarrow h \times (h - 1) = 20 \\ &\Rightarrow h^2 - h - 20 = 0 \\ &\Rightarrow (h - 5)\ (h + 4) = 0 \\ &\Rightarrow h = 5 \text{ or } h = -4 \end{aligned}$$

But, $h > 0$. Therefore, $h = 5$.

2. Given that

$$\begin{aligned} &2P(h, 2) + 98 = P(2h, 2) \\ &\Rightarrow 2 \times h \times (h - 1) + 98 = (2h) \times (2h - 1) \\ &\Rightarrow 2h^2 - 2h + 98 = 4h^2 - 2h \end{aligned}$$

$$\Rightarrow 98 = 2h^2$$
$$\Rightarrow h^2 = 49$$
$$\Rightarrow h = 7 \text{ or } h = -7$$

But, $h > 0$. Therefore, $h = 7$.

Type II: Permutations when all elements are distinct

Let m_1, m_2, m_3,..., m_k objects are of the first kind, second kind, third kind,..., k-th kind, respectively, and $m_1 + m_2 + m_3 + \cdots + m_k = h$. The number of permutations when m_1, m_2, m_3,..., m_k are taken at a time is [6].

$$P(h;\ m_1, m_2, m_3, \cdots, m_k) = \frac{h!}{m_1!\, m_2!\, m_3! \cdots m_k!}$$

Examples

1. $P(5;\ 1,\ 2,\ 2) = \frac{5!}{1!\,2!\,2!} = 30$
2. $P(4;\ 1,\ 1,\ 1, 1) = \frac{4!}{1!\,1!\,1!\,1!} = 24$

Example 7.14 Find the number of permutations that can be made by all letters of the words as follows.

(a) MISSISSIPPI (b) BOOKKEEPER (c) ATTITUDE

Solution (a) There are 11 letters in the given word in which S, I, P, M are distinct letters.

Letter	S	I	P	M
Frequency	4	4	2	1

Therefore, the required number of permutations

$$= P(11;\ 4, 4, 2, 1) = \frac{11!}{4!\,4!\,2!\,1!} = 34650.$$

(b) There are 10 letters in the given word in which B, O, K, E, P, R are distinct letters.

Letter	B	O	K	E	P	R
Frequency	1	2	2	3	1	1

Therefore, the required number of permutations $= P(10;\ 1, 2, 2, 3, 1, 1) = \frac{10!}{1!\,2!\,2!\,3!\,1!\,1!} = 151200$.

(a) There are 8 letters in the given word in which A,T, I, U, D, E are distinct letters.

Letter	A	T	I	U	D	E
Frequency	1	3	1	1	1	1

Therefore, the required number of permutations $= P(8;\ 1, 3, 1, 1, 1, 1) = \frac{8!}{1!\,3!\,1!\,1!\,1!\,1!} = 6720$.

Example 7.15 *How many 4 digit numbers are formed using the digits from the birthdate 10/01/5114.*

Solution There are 8 digits in the given birthdate in which 1, 0, 5, 4 are distinct digits.

Digits	1	0	5	4
Frequency	4	2	1	1

Therefore, the required number of permutations $= P(8;\ 4, 2, 1, 1) = \frac{8!}{4!\,2!\,1!\,1!} = 840$.

Type III: Permutations with repeated objects

The number of different permutations of h objects/elements taken m at a time, when every object is allowed to repeat any number of times is given by h^m [7, 8].

Examples

Example 7.16 *Prove that the number of rounded (circular) permutations of h dissimilar things is* $(h-1)!$.

Solution Let p be the required number of circular permutations on h distinct objects. If we break the circular permutations at unique position out of h positions, then this will become the linear permutation. Thus, for p circular permutations, there are $p \times h$ linear permutations. Therefore,

$p \times h = h! \quad \Rightarrow p = \frac{h!}{h} = (h-1)!$. Hence the proof.

Examples 7.17 *In how many ways 9 scientists of the Department Virology, World Health Organization, including Dr. Atharva and Dr. Omkar can sit in a circular table to discuss the health issues of people after COVID-19 pandemic such that Dr. Atharva and Dr. Omkar always sit together?*

Solution There are 10 scientists of the Virology department are sitting together for a meeting at a circular table. Dr. Atharva and Dr. Omkar are always sit together. So, both become a single unit. Therefore, there is an arrangement of 8 objects.

Thus, the 8 objects can be arranged at round table in $(8-1)! = 7! = 5040$ ways. The sitting of Dr. Atharva and Dr. Omkar can be arranged in $2!$ ways.

Therefore, the required number of ways $= 2! \times 7! = 10080$.

7.5 Combinations: Experiential Learning

(A) Brainstorming Experiment-III

Suppose, there are 14 students in the Department of Mathematics with 8 boys and 6 girl students. The Head of the department wants to constitute students' committee of 6 members for conducting co-curricular and extra-curricular activities more effectively. But, the condition is that there must be at least two girls' representatives and at least two boys' representatives. How many ways are there to constitute a committee?

(B) Inferences-III

We have to select 6 students out of 14 students. According to the conditions, there are the following three possible combinations of students in the committee.

i. 2 girls and 4 boys
ii. 3 girls and 3 boys
iii. 4 girls and 2 boys

We have to select 2 or 3 or 4 girls and 4 or 3 or 2 boys out of 6 girl representatives and 8 boy representatives, respectively. This is the selection procedure, so, it leads to the concept of the "COMBINATIONS".

Let us try to solve this practical problem by using the mathematical insights.

(C) Mathematical Insights-III

In some real life examples, we have to select some objects from independent objects. In such types of situations, we have to use the concept of Combinations for resolving the problems.

Definition The combination is the selection of some or all objects from the given set of objects.

Depending upon the nature of the selection, there two types of Combinations [5, 9].

Type I: Combinations when objects are different

The number of combinations of h different things taken m at a time, where $h \geq m$, is denoted by $C(h, m)$ or ${}^hC_m$ or $\binom{h}{m}$ and defined as ${}^hC_m = \frac{h!}{m!\,(h-m)!}$.

Properties:

We assume that $h \geq m$.

1. ${}^hC_m = \frac{h!}{m!\,(h-m)!} = \frac{{}^hP_m}{(h-m)!}$
2. ${}^hC_h = \frac{h!}{h!\,(h-h)!} = 1$
3. ${}^hC_0 = \frac{h!}{0!\,(h-0)!} = 1$

4. ${}^{h}C_m = \frac{h!}{m!(h-m)!} = \frac{h!}{(h-m)!(h-[h-m])!} = {}^{h}C_{h-m}$
5. ${}^{h}C_1 = \frac{h!}{1!(h-1)!} = h, \quad {}^{h}C_2 = \frac{h!}{2!(h-2)!} = \frac{h(h-1)}{2}, \quad {}^{h}C_3 = \frac{h!}{3!(h-3)!} = \frac{h(h-1)(h-2)}{6}$

Examples

Example 7.18 *Find the value of h if (i)* ${}^{h}C_2 = 45$*(ii)* ${}^{13}C_{h-2} = {}^{13}C_{2h-1}$.

Solution (i) We have

$$\begin{aligned}{}^{h}C_2 = 45 &\Rightarrow \frac{h!}{2!(h-2)!} = 45\\ &\Rightarrow h(h-1) = 90\\ &\Rightarrow h^2 - h - 90 = 0\\ &\Rightarrow (h-10)\ (h+9) = 0\\ &\Rightarrow h = 10 \text{ or } h = -9\end{aligned}$$

But, $h \geq 1$, Therefore, $h = 10$.

(ii) We have

$$\begin{aligned}&{}^{76}C_{h+2} = {}^{76}C_{2h-1}\\ &\Rightarrow \text{either } h+2 = 2h-1 \text{ or } h+2+(2h-1) = 76\\ &\Rightarrow h = 3 \text{ or } h = 25\end{aligned}$$

But, the result is not true for $h = 3$. Therefore, $h = 25$.

Example 7.19 *How many techniques (ways) can 16 late enrolled students be assigned to three practical batches P, Q, R if the batch P can accommodate 7 students, the batch Q can accommodate 5 students and the batch R can accommodate 4 students only.*

Solution The batches P, Q, and R can accommodate 7, 5, and 4 students, respectively.

The batch P can be assigned to 7 students in ${}^{16}C_7$ ways.
The batch Q can be assigned to 5 students in ${}^{9}C_5$ ways.
The batch R can be assigned to 4 students in ${}^{4}C_4$ ways.
Therefore, the required total number of ways is

$$\begin{aligned}&= {}^{16}C_7 \times {}^{9}C_5 \times {}^{4}C_4\\ &= \frac{16!}{7! \times 9!} \times \frac{9!}{5! \times 4!} \times 1 = \frac{16!}{7! \times 5! \times 4!}\end{aligned}$$

Example 7.20 *How many authorization number plates can be prepared if each plate comprises two unlike letters followed by three unlike digits? Explain the problem if the first digit is non-zero and the first letter is always H.*

Solution: There are five positions, two different letters and three different digits, in the number plate. We know that there are 26 letters and 10 digits.

Positions	1	2	3	4	5
Letter/Digit	Letter	Letter	Digit	Digit	Digit
The number of Ways	$^{26}C_1$	$^{25}C_1$	$^{10}C_1$	$^{9}C_1$	$^{8}C_1$

Therefore, the required number of ways is

$$= 26 \times 25 \times 10 \times 9 \times 8 = 468000$$

If the first digit is non-zero and the first letter is always H then consider the following table.

Positions	1	2	3	4	5
Letter/Digit	Letter	Letter	Digit	Digit	Digit
The number of Ways	1	$^{25}C_1$	$^{9}C_1$	$^{9}C_1$	$^{8}C_1$

Therefore, the required number of ways is $= 1 \times 25 \times 9 \times 9 \times 8 = 16200$.

Example 7.21 *How many lines can be drawn through 12 points on a circle? (Line passes through exactly two points).*

Solution As all points on the circle are not collinear. Therefore, the line passes through every pair of distinct points.

The total number of lines drawn through 12 points is

$$= {}^{12}C_2 = \frac{12!}{2! \times 10!} = 66$$

Type II: Combinations with Repetitions

The number of ways to fill s slots from c categories of objects with repetition is [9, 10].

$$C(c + s - 1, s) = C(c + s - 1, c - 1).$$

In this context,

$C(c + s - 1, s)$ = The number of non-negative integer solutions of $w_1 + w_2 + w_3 + \cdots + w_p = s$.

= The number of ways of placing s similar balls in c number of boxes.

= The number of binary c-1 one's and s zeros.

Examples

Example 7.22 *How many 4 combinations of* $\{aa, ba, ca, da, ea, fa\}$ *are there with repetitions?*

Solution Here, $h = 4$ and $p = 6$. The required number of 4 combinations from the given set is $C(p + h - 1, h) = C(9, 4) = 126$.

Example 7.23 *Find the number of techniques of placing 10 similar balls in 5 boxes.*

Solution The number of ways of placing $h = 10$ similar balls in $p = 5$ boxes is

$$C(p + h - 1, h) = C(14, 10) = 1001.$$

Example 7.24 *How many integral solutions are there to* $x_1 + x_2 + x_3 + \cdots + x_9 = 35$, where each $x_i \geq 2$?

Solution Let $y_i = x_i + 2, \therefore y_i \geq 0$. Therefore,

$$\begin{aligned} & x_1 + x_2 + x_3 + \cdots + x_9 = 35 \\ & \Rightarrow y_1 + 2 + y_2 + 2 + y_3 + 2 + \cdots + y_9 + 2 = 35 \\ & \Rightarrow y_1 + y_2 + y_3 + \cdots + y_9 = 35 - 18 = 17 \end{aligned}$$

The number of non-negative integral solutions are

$$C(c + s - 1, s) = C(9 + 17 - 1, 17) = C(25, 17) = 1081575$$

7.6 Case Studies and Projects

The following are some case studies and projects.

1. Suppose you want to constitute a committee of 4 boys and 4 girls' students from your department for conducting the co-curricular and extra-curricular activities of your department. How many ways are there for constitution of the committee? How many ways you can constitute a committee if a particular boy and girl are in the committee. How many ways are there if 2 members are present in every committee.
2. In a Cricket League, suppose different players having special expertise are there. Assuming the suitable data, form different cricket teams.
3. Portfolio assignment system in your institute
4. Selection of students' representative from every class and constitution of students' committees.
5. End-semester examination time table problem

6. Transportation problems (Selection of routes)

Exercise

1. Find the number of permutations that can be made all letters of the following words.

   i. ASSASSINATION
   ii. PIONEER
   iii. MATHEMATICS

2. How many four digits even numbers can be formed by using the 0, 1, 2, 4, 5, 8 when the repetition of digits are allowed?
3. A bank password involves three alphabets followed by three digits. How many different passwords are there?
4. How many different ways can 100 people be seated in a circular table?
5. Find the number of subsets of each possible size for a set containing 100 elements.
6. If m fair coins are tossed and the outcomes are recorded, then

   (a) How many sequences are possible?
   (b) How many sequences are there having 5 tails?
   (c) How many sequences are there having at least 100 heads?
   (d) How many sequences are there having at most 4 tails?

7. A fair six-sided die is tossed 5 times and the numbers appeared are noted in a sequence. How many different sequences are there?
8. Find the number of permutations of the of the letters in the following words.

   (a) BOOLEAN
   (b) DISCRETE
   (c) EXPERIENTIAL
   (d) STATISTICS

9. Find the number of non-negative integer solutions to $x + y + z + w = 30$.
10. Write any five real-life applications of permutations and combinations.

References

1. Aigner M (2023) Discrete mathematics. American Mathematical Society
2. Erciyes K (2021) Discrete mathematics and graph theory. Springer, Cham. https://doi.org/10.1007/978-3-030-61115-6
3. Kitaev S (2011) Patterns in permutations and words. Springer Science & Business Media. p 119. ISBN 978-3-642-17333-2
4. Weisstein EW. Permutation. mathworld.wolfram.com. Retrieved 2020-09-10

5. Bona M (2012) Combinatorics of permutations, 2nd edn. CRC Press. ISBN 978-1-4398-5051-0
6. Rosen K, Krithivasan K (2021) Discrete mathematics and its applications. MC Graw Hill India. ISBN-978930727353
7. Jiri M, Jaroslav N (2011) Invitation to discrete mathematics, 2nd edn. Oxford University Press. ISBN-9780198570424
8. Peter J (2000) Sets, logic and categories. Springer Publication. ISBN-9781852330569
9. Tremblay JP, Manohar R (2023) Discrete mathematical structures with applications to computer science. McGraw Hill Education. ISBN-9780074631133
10. Aigner M (2007) A course in enumeration. Springer GTM 238, pp 24–25. ISBN 978-3-540-39035-0

Chapter 8
Discrete Probabilities

The probability of success is difficult to estimate: but if we never search, the chance of success is zero.
—Philip Morrison

Abstract The probability of success is difficult to estimate; but if we never search, the chance of success is zero. Probability is an independent discipline of applied Mathematics, as it requires very basic Mathematics and it is developed for resolving numerous real-life problems from ancient times, as well as designing more pragmatic game strategies for winning the game. This chapter deals with the concepts of probability, discrete probability distributions and related results through the "BEIMI" perspective. Innovative and more pragmatic examples of real life define the concepts of discrete probability distributions. The probability of getting tails in consecutive 13 times in a cricket game motivates to define the concept of Poisson distribution. Additionally, innovative applications help readers to excel understanding of the concepts in meaningful ways. The chapter concludes with case studies and exercises aimed at connecting theoretical relational structures to real world applications across various disciplines.

Keywords Probability · Conditional probability · Probability distributions · Binomial · Poisson

8.1 Introduction

The probability is an independent discipline of applied Mathematics, as it requires very basic Mathematics and it is developed for resolving numerous real-life problems from ancient times, as well as designing more pragmatic game strategies for winning the game. In ancient times, people used probability theory without the Mathematical knowledge of it. People have been using the words "Chance" and "possibility" without knowing the Mathematical aspects. So, every common man is using the partial logic of probability theory in day-to-day activities. The following table depicts the development of the probability theory in Mathematical ways [1].

© The Author(s), under exclusive license to Springer Nature Singapore Pte Ltd. 2026

H. R. Bhapkar and P. N. Mahalle, *Discrete Mathematics: An Experiential Learning Perspective*, Studies in Autonomic, Data-driven and Industrial Computing,
https://doi.org/10.1007/978-981-95-7210-6_8

Year	Name of Mathematician	Development
1654	Blaise Pascal and Pierre Fermat	Solutions for winning the game
1657	Christian Huygens	De ratiociniis in ludo aleae
1713	James Bernoulli's	Ars conjectandi, probability into the mathematical theory
1805	Adrien Marie Legendre	Least squares method
1809–11	Laplace, Carl Friedrich Gauss	Probabilistic rationales for least squares
1812	Laplace	"Théorie analytique des probabilités," published in 1812

Many mathematicians and philosophers have contributed to the development of probability in different facets [2, 3].

This chapter deals with the basics of probability, properties, discrete probability distributions, and related examples.

8.2 Sample Space and Events: Experiential Learning

Let's consider the following brainstorming Experiment-I for understanding the concept of the Discrete Probability and related aspects.

(A) **Brainstorming Experiment-I**

Suppose there are 50 coins of equal size, shape, and weight. These coins are colored by two colors White and Black such that 25 coins are of each color. Now, we write odd numbers on the both sides of black coins and even numbers on the both sides of the white coins. Let H be the set of coins having labels 1 to 50 given by $H = \{1, 2, 3, \cdots, 50\}$. Let H_O be the set of coins having odd labels or black color and Let H_E be the set of coins having even labels or white color. Therefore,

$H_o = \{1, 3, 5, \cdots, 49\}$ and $H_E = \{2, 4, 6, \cdots, 50\}$.

There are 25 numbers in each set. Let H_P and H_C be the set prime numbers and composite numbers from 1 to 50. Therefore,

$$H_P = \{2, 3, 5, 7, 11, 13, 17, 19, 23, 29, 31, 37, 41, 43, 47\} \text{ and } H_C = H - H_P - \{1\}.$$

Hence, $|H_P| = 15$ and $|H_C| = 34$.

Let us consider the following sets.

Sr. No.	The numbers in H divisible by n	Required set H_n	$\|H_n\|$
1	$n = 5$	$H_5 = \{5, 10, 15, 20, 25, 30, 35, 40, 45, 50\}$	10
2	$n = 4$	$H_4 = \{4, 8, 12, 16, 20, 24, 28, 32, 36, 40, 44, 48\}$	14

(continued)

(continued)

Sr. No.	The numbers in H divisible by n	Required set H_n	$\|H_n\|$
3	$n = 9$	$H_9 = \{9, 18, 27, 36, 45\}$	5
4	$n = 16$	$H_{16} = \{16, 32, 48\}$	3
5	$n = 27$	$H_9 = \Phi$	0
6	$n = 6$ and $n = 4$	$H_{6\&4} = \{12, 24, 36, 48\}$	4
7	$n = 10$ or $n = 8$	$H_{10\ OR\ 8} =$ $\{8, 10, 16, 20, 24, 30, 36, 40, 48, 50\}$	10

There are many such combinations which will help us to understand the concepts of probability.

(B) **Inferences-I**

The above experiment helps to define the random experiment, outcome of an experiment, trials, sample space, events, types of events, equiprobable sample space, etc. [4].

(C) **Mathematical Insights-I**

1. **Random Experiment**: An experiment in which all results are recognized but uncertain, is called as a Random Experiment (RE). Getting any one coin from the above experiment, tossing a coin, rolling a die are the examples of the random experiment [5].
2. **Outcome**: The result of an experiment is known as the outcome of the experiment.
3. **Trial**: The trial is the performance of the random experiment once. Getting a coin from the set of 50 coins is a trial or the tossing of a coin each time is the trial.
4. **Sample Space**: It is the set of all imaginable results of the RE. It is denoted by H or SS.

In Experiment-I, the sample space is $H = \{1, 2, 3, \cdots, 50\}$.

5. **Event**: An event or case is any subset of the SS. It may contain the all elements of the sample space or no element. For example, in Experiment-I, getting a number divisible by 5, is an event. Getting two heads if two coins are tossed simultaneously is also an example of event.
6. **Exhaustive Events or Cases**: It is the set of all possible outcomes of a random experiment.
7. **Mutually Exclusive Events**: Two events P and Q are jointly exclusive events if the occurrence of one is independent of the occurrence of the other. In

Experiment-I, the occurrence of 5 and 13 are mutually exclusive events. In tossing a coin, the event of occurrence of H or T are mutually exclusive events.

8. **Equally Likely Events**: Two events P and Q are said to be equally likely events if every element has an equal chance of occurrence. In Experiment-I, getting a number 1, 2, 3, …, 50 are equally likely events [6].
9. **Independent Events**

   Two or more events are independent events if the existence of an event does not affect the occurrence of the remaining events. In Experiment-I, getting 13 in the first trial and getting 16 in the second trail, are independent events [7].
10. **Discrete Sample Space**:

    A sample space H is known as discrete if $|H| =$ Countable number. The sample space in the Experiment-I is discrete.
11. **Equiprobable Sample Space**: A sample space H is an equiprobable sample space if every element of a sample space has an equal chance of occurrence [6, 8].

Examples

Example 8.1 Example given in Experiment-I is random experiment. If getting a number divisible by 13 is an event then the sample space $H = \{1, 2, 3, \ldots, 50\}$ and event is $E_1 = \{13, 26, 39\}$. The second event is getting a number divisible by 5 then $E_2 = \{5, 10, 15, 20, 25, 30, 35, 40, 45, 50\}$. These two events are mutually exclusive.

Example 8.2 Statements corresponding to events and the corresponding set notations are depicted in the following table.

Sr. No.	Experiment	Sample space
1	Occurrence of an event E	E
2	Event in which E doesn't occur	$\overline{E}$
2	Events P and M occur	$P \cap M$
3	At least one of P or M occur	$P \cup M$
4	Neither I nor W occur	$\overline{I} \cap \overline{W} = \overline{(I \cup W)}$
5	Event I occurs or W occurs but not both	$(I \cap \overline{W}) \cup (\overline{I} \cap W) = I \oplus W$
6	I and W are mutually exclusive	$I \cap W = \varphi$
7	Events P, N, and M occur	$P \cup N \cup M$
8	Event P occurs but not N and M occur	$P \cap (\overline{N} \cap \overline{M})$
9	Event P doesn't occur but N and M occur	$\overline{P} \cap (N \cap M)$

8.3 Probability: Experiential Learning

(A) **Brainstorming Experiment-II**

Suppose, in the cricket world cup, there is a match between India and Australia. Now,

1. What is the possibility that India wins the toss?
2. What is the possibility that Australia wins the match?
3. What is the possibility of India to win the cricket world cup?

(B) **Inferences-II**

The above questions can be solved by the different definitions of the probability. The question 1 can be resolved by the classical definition of probability and questions 2 and 3 are solved by axiomatic definition of the probability [1, 9].

(C) **Mathematical Insights-II**

The probability can be defined in the following ways.

(I) **Mathematical or Classical Definition of Probability**:

Let S be an equiprobable finite sample space of a RE (Random Exp.) and E be any event of S, then the probability of occurrence of an event E is

$$P(E) = \frac{n\,(E)}{n\,(S)} = \frac{\text{Total number of favourable cases}}{\text{Total number of cases}}$$

Example 8.3 In Experiment-II, what is the possibility that India wins the toss?

Solution If we toss an unbiased coin once then the sample contains two elements H and T. To win the toss, India should get H outcome in this experiment. Therefore, the probability that India wins the toss is $P(E) = \frac{|E|}{|S|} = \frac{1}{2}$.

Note: We can't apply this definition for finding the answers of questions 2 and 3 as the corresponding sample space is not equiprobable sample space. The strengths of both the teams are different [9].

(II) **Axiomatic Definition of Probability**: Let S be not an equiprobable sample space (ESS) of a RE and E be any event of a SS then the probability of occurrence of an event E is a real number satisfying the following axioms:

i. $0 \leq P(E_i) \leq 1, \quad \forall\, i = 1, 2, \ldots, n$

ii. $P(S) = 1 = \sum_{\forall\, i} P(E_i)$

Example 8.4 In a class of 60 students, what would be the probability of a student to secure the first rank in the End Semester Examination?

Solution The answer of this question is not 1/60 as the intelligence and resources of all students are different. This is not an ESS. Let P_c be the probability of the c-th student of the class by considering their previous all records such that.

i. $0 \leq P_i \leq 1$, for $i = 1, 2, 3, \ldots, 60$
ii. $\sum_{i=1}^{60} P_i = 1$

The probability of a clever student may be more than other students.

Theorems on Probability

Theorem 8.1 *Show that* $P(\varphi) = 0$.

Proof The events S (Sample space) and φ are mutually exclusive. Therefore,

$$
\begin{aligned}
&P(S \cup \varphi) = P(S) + P(\varphi) \\
&\Rightarrow P(S) = P(S) + P(\varphi) \\
&\Rightarrow P(\varphi) = 0
\end{aligned}
$$

1. For two sets I and W.

$$P(I \cup W) = P(I) + P(W) - P(I \cap W).$$

2. For three sets H, M, T,

$$
\begin{aligned}
P(H \cup M \cup T) = P(H) + P(M) + P(T) - P(H \cap M) \\
- P(H \cap T) - P(M \cap T) + P(H \cap M \cap T).
\end{aligned}
$$

Examples

Example 8.5 What is the probability that a year 2028 will contain 53 Mondays?

Solution Required probability = $\frac{2}{7}$.

Example 8.6 If 9 persons take seats at random at a circular table, find the probability that 2 particular brigadiers are seated next to each other.

Solution Nine persons can be settled on a round table in (9-1)! = 8! Ways.

If two persons are to be seated together, then we may regard them to have been "tied" up, so that now the seating can take place in (8-1)! = 7! Ways. But these two persons can also seat themselves together in 2 ways.

$$P = \frac{\text{Favourable ways}}{\text{Total number of ways}} = \frac{7!2}{8!} = \frac{1}{4}$$

Example 8.7 A problem of DM is given to 5 students U, V, W, R, and T. Their probabilities of solving the problem are1/2, 1/3, 1/ 4, 1/5, and 1/6, respectively. What is the probability that problem get solved?

Solution Required Probability = probability that at least one of the student solve it.

$$P(U) = \frac{1}{2} \Rightarrow P(\overline{U}) = 1 - \frac{1}{2} = \frac{1}{2}$$
$$P(V) = \frac{1}{3} \Rightarrow P(\bar{V}) = 1 - \frac{1}{3} = \frac{2}{3}$$
$$P(W) = \frac{1}{4} \Rightarrow P(\bar{W}) = 1 - \frac{1}{4} = \frac{3}{4}$$

$$P(R) = \frac{1}{5} \Rightarrow P(\bar{R}) = 1 - \frac{1}{5} = \frac{4}{5}$$
$$P(T) = \frac{1}{6} \Rightarrow P(\bar{T}) = 1 - \frac{1}{6} = \frac{5}{6}$$

$\therefore$ Probability that problem is not solved

$$= P(\overline{U}) \times P(\overline{V}) \times P(\overline{W}) \times P(\overline{R}) \times P(\overline{T})$$

Probability that problem is solved

$$= 1 - P(\overline{U}) \times P(\overline{V}) \times P(\overline{W}) \times P(\overline{R}) \times P(\overline{T})$$
$$= 1 - \frac{1}{6} = \frac{5}{6}$$

8.4 Conditional Probability: Experiential Learning

(A) **Brainstorming Experiment-III**

In Experiment-II, there are two main and interesting questions which motivates us to think the probability concept more innovative ways.

1. What is the possibility that India wins the game if it is played in India?
2. What is the probability that India wins the match if it is played in Australia?

(B) **Inferences-III**

The answers of both the questions are different, as per the previous records of the matches and the concerned locations. This philosophy leads the concept of "CONDITIONAL PROBABILITY".

(C) **Mathematical Insights-III**

The chance of the presence of the event, when the other event has been already happened, is called the conditional probability. The probability of an event E given G is denoted as $P(E|G)$ [1, 6].

Theorem 1 *For two events E and G*

$$P(E \cap G) = P(G)\,P(E|G)$$

Notes:

1. If H and M are independent then $P(H \cap M) = P(H) \times P(M)$.
2. If H and M are independent events then H and $\overline{M}$, $\overline{H}$ and M, $\overline{H}$ and $\overline{M}$ are independent events. That is $P(\overline{H} \cap \overline{M}) = P(\overline{H}) \times P(\overline{M})$.

Theorem 8.2 *Theorem on Total Probability*

If $H_1, H_2, H_3, \ldots, H_n$*be mutually exclusive events with* $P(H_i) \neq 0, \quad \forall i = 1, 2, 3, \ldots, n$*then for any event* $W \subseteq \bigcup_{i=1}^{n} H_i$*, prove that*

$$P(W) = \sum_{i=1}^{n} P(H_i) \times P(W|H_i).$$

Baye's Theorem

If $H_1, H_2, H_3, \ldots, H_n$*be mutually exclusive events with* $P(H_i) \neq 0, \quad \forall i = 1, 2, 3, \ldots, n$*then for any event* $W \subseteq \bigcup_{i=1}^{n} H_i$*, such that* $P(W) > 0$*, prove that*

$$P(H_i|W) = \frac{P(H_i) \times P(W|H_i)}{\sum_{i=1}^{n} P(H_i) \times P(W|H_i)}.$$

Examples

Example 8.8 A software company selects marketing professionals on the basis of an aptitude test. Past experience indicates that only 75% of the candidates were found satisfactory in actual marketing, 80% had passed the aptitude test. Only 20% of those found unsatisfactory has passed the test. Given that the candidate passed the test what is the probability that he would be found satisfactory.

Solution Let H be the event that the candidate passed the aptitude test and M be the event that he would be found satisfactory. Find P (M/H). Consider M and $\overline{M}$

$$P(M/H) = \frac{P(H/M) \cdot P(H/M)}{P(H/M) \cdot P(M) + P(H/\overline{M})P(H/\overline{M})}$$

$$P(M) = 0.75, \quad P(\overline{M}) = 0.25$$

$$P(H/M) = 0.80, \quad P(H/\overline{M}) = 0.25$$

$$P(M/H) = \frac{0.75 \times 0.8}{(0.75)\ (0.8) + (0.25)\ (0.2)} = 0.923.$$

8.5 Probability Distributions: Experiential Learning

In real life, there are numerous situations where it might be difficult and time-consuming task to find all elements of the sample space or a particular event. There are some patterns in the experiment or common attributes which could be solved by the theory of different probability distributions. Let us consider the basic definitions of required for the deepen understanding of this concept.

1. **Random Variable (RV)**: It is a function from the set of all feasible results of a random experiment to the set of real numbers.
2. **Types of the Random Variables**: There are three types of RV.
a. Discrete RV: It takes only finite or countably infinite number of values.
b. Continuous RV: It takes all possible values in an interval.
c. Mixed RV: It takes discrete and continuous values.
3. **Probability Mass Function (PMF)**: The function $P(w = w_i) = P(w_i)$ is called a PMF of the RV, W if

    i. $P(w_i) \geq 0, \quad$ for every i
    ii. $\sum_{i=1}^{\infty} P(w_i)$

4. **Cumulative Distribution Function (CDF)**: The cumulative distribution function or distribution function of RV, W is defined as

$$F_w(w) = P(W \leq w) = \sum_{w_i \leq w} P(w_i), \text{ where } P(w_i) \text{ is the PMF of } W.$$

Properties

i. $0 \leq F(w) \leq 1$,
ii. $F(\infty) = 1$ and $F(-\infty) = 0$,
iii. $P(h < w \leq p) = F(p) - F(h)$,
iv. $P(h < w < p) = F(p) - F(h) - P(w = p)$,
v. $F(h) \leq F(m)$ if $\ h < m$.

5. Mathematical Expectation: The expected value of a random variable W is defined as $E(W) = \sum_{i=1}^{n} w_i P(w_i)$, provided the series is absolutely convergent.

    Properties

    i. If c is a constant then $E(c) = c$.
    ii. $E(aW + bU) = aE(W) + bE(U)$, where W and U are two RV and a, b are constants.
    iii. If W and U are independent RV then $E(WU) = E(W)E(U)$.

6. Moments of a random variable: The r-th moment about the origin of a discrete random variable W is defined as $\mu_r^1 = E(W^r)$.

    The r-th moment about the mean μ of a discrete random variable W is defined as $\mu_r = E[(W - \mu)^r]$.

7. **Moment Generating Function (MGF)**: The MGF of a RV, W about the origin is defined as

$$M_W(t) = E(e^{tW}) = \sum_{x_i} e^{t\,W_i} P(w_i)$$

where, $P(w_i)$ is the pmf of DRV.

8.6 Binomial Distribution: Experiential Learning

(A) **Brainstorming Experiment-IV**

Let us consider the real situation in the cricket history. Recently, Mr. Rohit Sharma, Captain, Indian Cricket Team, lost 12 consecutive tosses and made a world record. Now, understand the mathematical perspective of this situation. If a coin is tossed 12 times, then what is the probability of receiving 12 times consecutive Tails? It is very difficult and a time taking process to find all the elements of the SS. There are 2^{12} = 4096 possible outcomes. But, in this experiment, there is a hidden pattern. There are only two possible outcomes for each trial.

(B) **Inferences-IV**

The above experiment and the situation motivates us to define a new theory in which there are only two possible outcomes. That is, there are only two nominees and the finite number of trials.

This philosophy leads to the concept of "BINOMIAL DISTRIBUTION."

(C) **Mathematical Insight-IV**

If we toss a coin, every time there are two mutually exclusive possible outcomes H and T. Every trial is independent of the previous trials. Let us perform n trials of this experiment.

Let s = probability of getting success and d = the probability of getting failure or defeat. As there are only two outcomes, $s + d = 1$ [2, 9].

The probability of getting k successes in n trials is given as follows.

$$\begin{aligned}
&P(k \text{ successes in } n \text{ trials})\\
&= P[(S \times S \times S \times \cdots \times S)_{k \text{ times}} (F \times F \times F \times \cdots \times F)_{n-k \text{ times}}]\\
&= [P(S) \times P(S) \times P(S) \times \cdots \times P(S)]_{k \text{ times}} [P(F) \times P(F) \times P(F) \times \cdots \times P(F)]_{n-k \text{ times}}\\
&= [s \times s \times s \times \cdots \times s]_{k \text{ times}} [d \times d \times d \times \cdots \times d]_{n-k \text{ times}}\\
&= s^k d^{n-k}
\end{aligned}$$

But, the k successes and n-k failures can occur in ${}^nC_k$ mutually exclusive ways. Therefore,

$P(k \text{ successes in } n \text{ trials}) = {}^nC_k\, s^k d^{n-k}$ where, $k = 0, 1, 2, 3, \ldots, n$

Now, substituting the different values of $k = 0, 1, 2, 3, \ldots, n$, we get the following formulae.

k	0	1	2	3	…	$n-1$	n
$P(k)$	${}^nC_0\ s^0 d^{n-0}$ $= d^n$	${}^nC_1\ s^1 d^{n-1}$ $= nsd^{n-1}$	${}^nC_2\ s^2 d^{n-2}$	${}^nC_3\ s^3 d^{n-3}$	…	${}^nC_{n-1}\ s^{n-1} d^1$	${}^nC_n\ s^n d^0$

Assumptions of Binomial Distribution

i. The experiment consists of finite number of repeated trials.
ii. Every trial has exactly two independent outcomes, say Head or Tail, Pass or Fail, etc.
iii. Repeated trials are independent.

Binomial Distribution

Definition A discrete RV, W is said to follow Binomial distribution if its pmf is defined as

$$P(k) = {}^nC_k s^k d^{n-k} \text{ for, } k = 0, 1, 2, 3, \ldots, n,$$

where n is the positive integer and $0 < s < 1$, $s + d = 1$ [1, 6].

Alternately, the probability of getting k successes in n independent trial is $P(k) = {}^nC_k\ s^k d^{n-k}$.

Mean and Variance of the Binomial Distribution:

(1) Mean

$$P(k) = {}^nC_k s^k d^{n-k}$$

$$\begin{aligned}
\text{Mean } \mu &= \sum_{k=0}^{n} k \cdot P(k) = \sum_{k=0}^{n} k \cdot {}^nC_k s^k d^{n-k} \\
&= 0 + 1 \cdot {}^nC_1 s d^{n-1} + 2 \cdot {}^nC_2 s^2 d^{n-2} + 3 \cdot {}^nC_3 s^3 d^{n-3} + \cdots + n \cdot {}^nC_n s^n d^0 \\
&= ns\left[d^{n-1} + (n-1)d^{n-2} \cdot s + \frac{(n-1)\,(n-2)}{2 \cdot 1} d^{n-3} \cdot s^2 + \cdots + (n-1)C_{(n-1)} s^{n-1}\right] \\
&= ns(d+s)^{n-1}
\end{aligned}$$

$$\therefore \mu = ns \quad \text{as } s + d = 1$$

Mean of binomial distribution is ns.

(2) Variance

$$\begin{aligned}\sigma^2 &= \sum_{k=0}^{n} k^2 P(k) - \mu^2 \\ &= \sum_{k=0}^{n} (k + k^2 - k) \cdot P(k) - \mu^2 \\ &= \sum_{k=0}^{n} k \cdot P(k) + \sum_{k=0}^{n} k(k-1)P(k) - \mu^2 \\ &= \mu + n(n-1)s^2[d+s]^{n-2} - \mu^2\end{aligned}$$

$$\text{as} \quad s + d = 1 \quad \text{and} \quad \mu = ns$$

$$\begin{aligned}\text{Variance} \quad \sigma^2 &= ns + n(n-1)s^2 - n^2 s^2 \\ &= ns[1 + (n-1)s - ns] \\ &= ns[1 + ns - s - ns] = ns(1-s)\end{aligned}$$

$$\textit{Variance}\ \sigma^2 = nsd$$

Therefore, the standard deviation of the binomial distribution is $\sigma = +\sqrt{nsd}$.

Example 8.9 A dice is thrown 5 times. If getting numbers 1 or 3 or 5 is a success, what is the probability of getting 4 successes.

Solution Odd numbers are 1, 3, 5

$$\text{Probability of an odd} = \frac{3}{6} = \frac{1}{2} \text{i.e. } p = \frac{1}{2}$$

$$\text{Probability of not an odd} = 1 - \frac{1}{2} = \frac{1}{2} \text{i.e. } q = \frac{1}{2}$$

$$P(k) = {}^{n}C_k s^k d^{n-k}$$

$$P(4) = {}^{5}C_4 \left(\frac{1}{2}\right)^4 \left(\frac{1}{2}\right)^{5-4} + {}^{5}C_5 \left(\frac{1}{2}\right)^5 \left(\frac{1}{2}\right)^{5-5}$$

$$= \frac{5}{32} + \frac{1}{32} = \frac{6}{32}.$$

Example 8.10 Probability of a soldier striking a goal is1/4. If the soldier fires seven times, what is the probability of striking the goal at least twice?

$$\text{Solution} \quad \text{Here} \quad s = \frac{1}{4}, d = \frac{3}{4}, n = 7$$

$$P(k) = {}^{n}C_k (s)^k (d)^{n-k}$$

$$p(\text{at least twice}) = p(k \geq 2)$$
$$= P(2) + P(3) + P(4) + P(5) + P(6) + P(7)$$

As the sum of probability = 1
i.e., $P(0) + P(1) + \ldots + P(7) = 1$

$$\text{P}(2) + \text{P}(3) + \text{P}(4) + \text{P}(5) + \text{P}(6) + \text{P}(7) = 1 - \text{P}(0) - \text{P}(1)$$
$$= 1 - \left[{}^7C_0\left(\frac{1}{4}\right)^0\left(\frac{3}{4}\right)^7\right] - \left[{}^7C_1\left(\frac{1}{4}\right)^1\left(\frac{3}{4}\right)^6\right]$$
$$= \frac{9094}{16384} = 0.555$$

Example 8.11 If Mean = 6 and variance = 2, in binomial distribution then.

Find $P(k \geq 1)$

Solution Mean = ns =6
Variance = nsd = 2

$$6 \times d = 2 \Rightarrow d = \frac{1}{3} \text{ and } s = 1 - \frac{1}{3} = \frac{2}{3}$$
$$\text{As} \quad ns = 6 \Rightarrow n \cdot \frac{2}{3} = 6 \Rightarrow n = 9$$

$$P(k \geq 1) = 1 - \left\{{}^nC_0 p^r \cdot q^{n-r}\right\}$$
$$= 1 - \left[{}^9C_0\left(\frac{2}{3}\right)^0 \times \left(\frac{1}{3}\right)^9\right] = 0.9999.$$

8.7 Poisson Distribution: Experiential Learning

(A) Brainstorming Experiment-V

During the COVID-19 pandemic, a new type of vaccine came to the market, and many people were injected with that vaccine. However, that vaccine was not 100% safe. It was observed that the probability that an individual suffers a bad response from the vaccine is 0.01. The vaccine was given to more than 100 crore people. In our city, fifteen thousand people got that vaccine. The district collector of the city wanted to know a detailed report about the people who were suffering from bad reactions.

Now, the main questions are, how many people will be suffering from bad reactions, and how to determine the probability out of 15,000 people, exactly 200 people will suffer bad reactions from the vaccine.

(B) Inferences-V

In this experiment, there are only two possibilities: either people suffer bad reactions from the vaccine or not. Here, the probability of suffering the bad reactions from the vaccine is a very less, and the number of people is a very big number for calculations. Therefore, we can't apply the Binomial distribution in this case. Thus, we require a new distribution that deals with two outcomes only and large n.

(C) Mathematical Insights-V

When "s" the probability of success is very small and "n" the number of trials is very large and ns is finite, we get another distribution called "Poisson Distribution." It is considered as limiting case of Binomial distribution with $n \to \infty$, $s \to 0$ and ns remains finite.

Consider the Binomial distribution

$$B(n, s, k) = {}^{n}C_{k} s^{k} d^{n-k}$$
$$= \frac{n(n-1)\,(n-2)\cdots(n-k+1)}{k!}\, s^{k}(1-s)^{n-k}$$
$$\text{Let } z = ns \quad \therefore\ ps = \frac{z}{n}$$
$$B(n, s, k) = \frac{ns\,(ns-s)\,(ns-2s)\cdots(ns-(k-1)s)}{k!} \times \frac{\left(1-\frac{z}{n}\right)^{n}}{\left(1-\frac{z}{n}\right)^{k}}$$

$$\text{Taking the } \lim_{s\to 0} \text{ it as } n \to \infty \text{ and } ns = z$$

$$\text{Lim } B(n, s, k) = \frac{z^{r} \cdot e^{-z}}{k!}\left[\lim_{n\to\infty}\left(1-\frac{z}{n}\right)^{n} = e^{-z}\right].$$

The distribution with frequencies given by $P(k) = \sum_{k=0}^{\infty} \frac{e^{-z}.z^{k}}{k!}$ corresponding to 0, 1, 2, ... k, success is called Poisson distribution [3, 5].

Mean and Variance

$$\text{we have } P(k) = \frac{e^{-z}z^{k}}{k!}$$
$$\text{Mean}\quad \mu = \sum_{k=0}^{\infty} k \cdot P(k) = \sum_{k=0}^{\infty} k \cdot \frac{e^{-z}z^{k}}{k!}$$
$$= e^{-z}\sum_{k=1}^{\infty}\frac{z^{k}}{(k-1)!}$$
$$= e^{-z}\left[z + \frac{z^{2}}{1!} + \frac{z^{3}}{2!} + \cdots\right]$$
$$= z \cdot e^{-z} \cdot e^{z} = z.$$

$$\text{Mean } = \mu = \text{z} = \text{ns}$$

$$\begin{aligned}
\text{Varience } \sigma^2 &= \sum_{k=0}^{\infty} k^2 P(k) - \mu^2 \\
&= \sum_{k=0}^{\infty} k^2 \frac{e^{-z} z^k}{k!} - z^2 \\
&= e^{-z}\left[\frac{1^2 \cdot z}{1!} + \frac{2^2 \cdot z^2}{2!} + \frac{3^2 \cdot z^3}{3!} + \cdots\right] - z^2 \\
&= z \cdot e^{-z}\left[1 + \frac{2z^2}{1!} + \frac{3z^3}{2!} + \cdots\right] - z^2 \\
&= z(1+z) - z^2 = z \\
\text{Variance } \sigma^2 &= \lambda = ns. \\
\text{Mean} &= \text{variance} = ns.
\end{aligned}$$

Recurrence Formula

$$P(k+1) = \frac{\lambda}{k+1} P(k)$$
$$k = 0, 1, 2, 3$$

Example 8.12 1% of tokens from a certain engine are faulty. What is the probability of (i) No faulty (ii) One faulty (iii) Two faulty (iv) Two or more faulty in a sample of 100?

Solution Here $p = 1/100$, $n = 100$, $np = z = 1$.

Thus, the Poisson's distribution takes form $\sum_{r=0}^{\infty} e^{-1} \cdot \frac{1}{r!}$.

Thus, (i) The probability of no faulty = $\frac{e^{-1}}{0\,!} = 0.36791$.

(ii) The probability of one faulty = $\frac{e^{-1}}{1\,!} = e^{-1} = 0.3679$

(iii) The probability of two faulty = $\frac{e^{-1}}{2\,!} = 0.1840$

(iv) The probability of two or more faulty

$$\begin{aligned}
&= \frac{e^{-1}}{2!} + \frac{e^{-1}}{3!} + \frac{e^{-1}}{4!} + \cdots = e^{-1}\left\{\frac{1}{2!} + \frac{1}{3!} + \frac{1}{4!} + \cdots\right\} \\
&= e^{-1}\left[\left(1 + \frac{1}{2!} + \frac{1}{3!} + \frac{1}{4!} + \cdots\right) - 1 - 1\right] \\
&= e^{-1}\{e - 2\} = 0.3679 \times 0.7183 = 0.2642
\end{aligned}$$

8.8 Case Studies and Projects

Consider the following case studies/projects based on the topics of this chapter.

1. What is the probability of a particular student of your class/division to stand first in the end-of-semester examination?
2. Write a probabilistic model to find the probability of a particular student to passing all courses of this academic year.
3. By visiting any industry in the vicinity of your institution and get desirable data, and find the probability of getting defective articles manufactured by the industry.
4. Suppose you are appearing for an interview for the post of Junior Scientist. There are 2 vacancies and 10 candidates have appeared for the interview. Assuming the suitable data, what is the probability of your selection?
5. Poisson Distribution and its applications.
6. Binomial Distribution and its applications.
7. Bayes' Theorem and its applications

Exercise

1. What is the probability that at least two students in a class of 30 students have the same birthday?
2. A problem of discrete mathematics is given to five students P, Q, R, S, T and their chances of solving it are 0.5, 0.09, 0.6, 0.55, 0.75, respectively.

   i. What is the probability that the problem is resolved?
   ii. What is the probability that at least two students can solve the problem?
   iii. What is the probability that at most three students can solve the problem?

3. Find the probability of drawing (i) a card of spades (ii) a king, (iii) a king or queen, (iv) card having number five on it, from a pack of cards.
4. Three bags P, Q, R involve 4 white, 6 red and 5 white, 5 red and 3 white, 7 red balls, respectively. One of the bag is selected at random and a ball is drawn from it. If the ball drawn is white. What is the probability that it is drawn from the second bag?
5. A pair of dice is thrown 10 times. If getting a doublet (same numbers) is considered as success, find the probability of getting

   i. 4 successes
   ii. 1 successes
   iii. At least 3 successes
   iv. At most 4 successes.

References

1. Pearson ES, Kendall MG (eds) (1970) Studies in the history of statistics and probability, vol 1. Darien, Connecticut (Hafner)

2. Shafer G (1988) The St. Petersburg paradox. In: Kotz S, Johnson NL (eds) Encyclopedia of statistical sciences, vol 8. Wiley, New York, pp 865–870
3. Viti A, Terzi A, Bertolaccini L (2015) A practical overview on probability distributions. J Thorac Dis 7(3):E7–E10. https://doi.org/10.3978/j.issn.2072-1439.2015.01.37
4. Aigner M (2023) Discrete mathematics. American Mathematical Society
5. Erciyes K (2021) Discrete mathematics and graph theory. Springer Cham. https://doi.org/10.1007/978-3-030-61115-6
6. Rosen K, Krithivasan K (2021) Discrete mathematics and its applications. MC Graw Hill India. ISBN-978930727353
7. Jiri M, Jaroslav N (2011) Invitation to discrete mathematics, 2nd edn. Oxford University Press. ISBN-9780198570424
8. Peter J (2000) Sets, logic and categories. Springer Publication. ISBN-9781852330569
9. Tremblay JP, Manohar R (2023) Discrete mathematical structures with applications to computer science. McGraw Hill Education. ISBN-9780074631133

Chapter 9
Graphs

Graphs stand or fall by their choice of nodes and edges.
—Watts & Strogatz"

Abstract Graph Theory is an independent discipline of Mathematics that connects many disciplines of Mathematics together in more pragmatic approaches. The novel work of graph theory was inspired by unwavering efforts to fathom and resolve day-to-day problems in a more logical perspective. Graphs stand or fall by their choice of nodes and edges. This chapter deals with the concepts of Graphs, types of graphs, operations of graphs, isomorphism of graphs, connectivity of graphs, coloring of graph attributes, matrix representation of graphs and various algorithms for finding the shortest path. This chapter follows the concept of the "BEIMI" to define various concepts of graphs and find out the associated applications. The chapter culminates in case studies, projects and pedagogical exercises aimed at reconciling theoretical relational frameworks with their multidisciplinary practical implementations.

Keywords Graphs · Isomorphic graphs · Connectivity · Coloring of graphs · Matrix representations · Algorithms

9.1 Introduction

The graph theory is an independent discipline of Mathematics that connects many disciplines of Mathematics together in more pragmatic methods. The novel work of the graph theory was inspired by unwavering efforts to fathom and resolve day-to-day problems in a more pragmatic perspective. In 1735, to solve the famous Konigsberg bridge problem, a great Mathematician, Euler used graphical representations for the first time to replicate the situation. He solved this problem, and used graphs to solve numerous problems of different kinds. Euler is the father of the graph theory due to his significant contributions to the development of the graph theory [1, 2].

Many great mathematicians namely, Mobius, Gaustar Kirchhoff, De-morgan William Hamilton, Alfred Kempe, P. K. Tait, Narsing Deo, etc., have contributed

© The Author(s), under exclusive license to Springer Nature Singapore Pte Ltd. 2026
H. R. Bhapkar and P. N. Mahalle, *Discrete Mathematics: An Experiential Learning Perspective*, Studies in Autonomic, Data-driven and Industrial Computing,
https://doi.org/10.1007/978-981-95-7210-6_9

many theories, results, and applications in the development of the graph theory. There are numerous causes for the acceleration of special curiosity in graph theory because it has thousands of applications in computer engineering, communication science, electrical and civil engineering, cryptography, operations research, applied sciences, genetics, pandemics, and many more. The role of the graph theory is crucial in controlling the COVID-19 pandemic. The name of the pandemic "COVID-19" is derived from the Corona product of graphs, the quarantine concept originated from the disconnected sets of the graph [3–6]. As there are significant and notable applications of graph theory, it has become an attraction for students and researchers [1].

This chapter deals with the definition of graphs, their types, operations on graphs, isomorphism of graphs, connectivity, graph coloring, shortest path algorithms, and case studies in the graph theory.

9.2 Graphs and Multigraphs: Experiential Learning

(A) **Brainstorming Experiment-I**

Suppose, there are five main cities in a state that have been contributing significantly for the development of the state and in the country also. These cities are well connected with each other as well as with the other cities of the country in all possible ways. There are many roadways connected to different areas of the city. Let us denote the city by points or small circles and the roads connecting them by a continuous arc. Now, observe the possible roadways among these cities and represent it by graphs.

(B) **Inferences-I**

The following are the special observations of the above Experiment-I.

1. All vertices many be connected to each other (Graph).
2. There may be a city X that has only one road with city Y, and no road with other cities (Pendent Vertex and Pendent Edge).
3. There may be two different roads joining two cities (Multiple or Parallel Edges).
4. There may be a one road that joins the city X to itself (Loop).

(C) **Mathematical Insights-I**
(I) **Graphs**

A graph H involves two components $V(H)$ and $V(E)$ such that

1. $V(H) = \{h_1, h_2, h_3, \ldots, h_n\}$ is the set of elements known as vertices or nodes or points. The number of vertices n may be finite or infinite.
2. $E(H)$ is the set of unordered pair of vertices, known as the edge set. The elements of the edge set are called as edges or arc or branches of H.

The above structure can be represented diagrammatically in the plane, so it is called as graph.

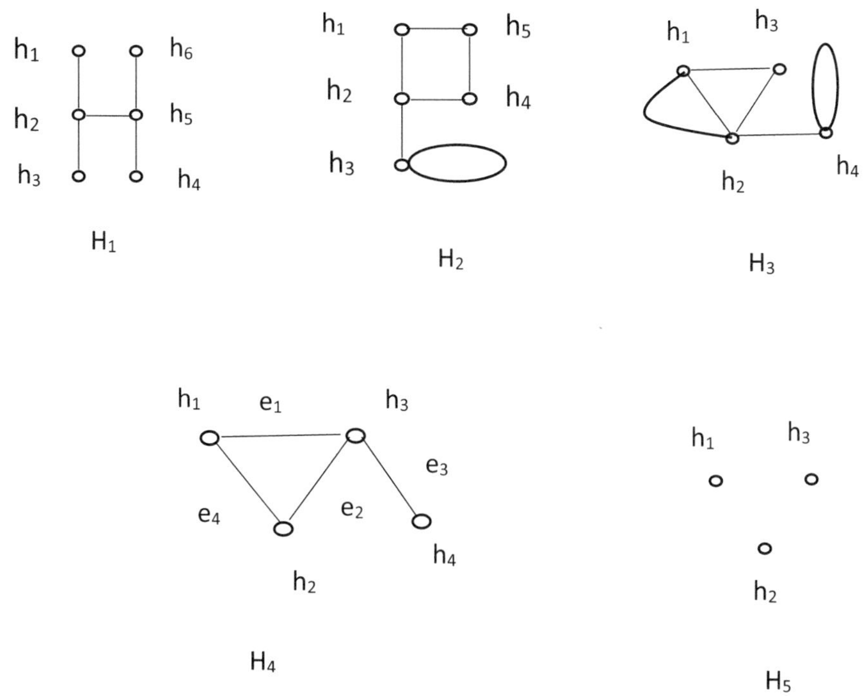

Fig. 9.1 Basic graphs

It is denoted by H or $H\ (V, E)$.

In the applications or everyday illustrations, the vertex set could be the collection of people or cities or states or animals or objects, etc. All edges are symbolized by an unbroken arc or straight line. The edges may be the relation among objects or routes among states or cities.

Consider the following graphs shown in Fig. 9.1.

i. If an edge $e = \{h_i,\ h_j\}$ joins vertices h_i and h_j then we can say that the edge e is incident at the vertices $h_i\ \&h_j$.
ii. Two vertices $h_i\&h_j$ are said to be adjacent if $\exists$ an edge $e = \{h_i,\ h_j\}$ in the graph H. In graph $H2$, the vertex h_1 is adjacent with the vertices h_2 and h_5.
iii. Two edges $e_a = \{h_i,\ h_j\}$ and $e_b = \{h_s,\ h_t\}$ are said to be adjacent iff $h_i = h_s$ or $h_j = h_t$.
iv. Two edges are called parallel or multiple edges iff they have the common end vertices, i.e., $e_a = \{h_i,\ h_j\} = e_b$. In graph H3, the vertices h_1 and h_2 are join by two edges. These edges are parallel edges.
v. An edge $e_a = \{h_i,\ h_j\}$ is called a loop iff $h_i = h_j$. In graph $H3$, $e = \{h_4, h_4\}$ is a loop.

(II) **Multigraphs**

1. A graph having multiple edges is called a multigraph. The graph $H3$ is a multigraph as it has a multiple edges and the remaining graphs are not multigraphs [1].
2. A graph having at least one loop but not multiple edges, is called a Pseudograph. The graph $H2$ is a Pseudograph but all remaining graphs are not Pseudographs.
3. A graph without loops and parallel edges, is called a simple graph. The graphs $H1$, $H4$, and $H5$ are simple but H_2 and H_3 are not simple graphs.
4. A graph H is called a null graph (N_n) if $E(H) = \varphi$. The graph $H5$ is a null graph on 3 vertices.
5. A graph H is called a finite graph iff $|E(H)|$ & $|V(H)|$ are finite sets. Otherwise, the graph is infinite.
6. Directed graph is a graph in which every edge is a directed edge. The directed edge e from the vertex h_i to h_j is represented as $e = (h_i,\ h_j)$. Here, $(h_i,\ h_j) \neq (h_j,\ h_i)$. The non-directed graphs are called as graphs only.
7. Order and Size of graph:

   Order $(H) = |V(H)|$ and size $(H) = |E(H)|$

   The null graph is $N(n, 0)$. The size of the graph $H1$ is (6, 5).
8. Degree of a Vertex

$$d\ (w) = |\{e\ /\ e = \{w, m\},\ \forall m \in V(H)\}|.$$

   That is, $d(w)$ is the number edges incident at vertex w in the graph H. A loop edge contribute degree 2 at that vertex.

   The degree of the graph is defined as

$$D(H) = \sum_{\forall w \in V(H)} d(w)$$

   Note: Every edge contribute degree 2 in $D(H)$. So, although the number of vertices are even or odd the $D(H)$ is always even.

   In graph $H2$, $d\ (h_1) = d\ (h_4) = d\ (h_5) = 2$, $d\ (h_2) = 3$, and $d\ (h_3) = 3$. Therefore, $D\ (H_2) = 12$.
9. Degree sequence of a graph: If $V(H) = \{h_1, h_2, h_3, \ldots, h_n\}$ then the degree sequence of the graph H is $(d_1, d_2, d_3, \ldots, d_n)$, where $d_i = d(h_i)$, $\forall\ i = 1, 2, 3, \ldots, n$.

   Two graphs U and V are said to be degree equivalent if their degree sequences are same.

Theorem 9.1 *Handshaking Lemma: Let H (V,E)be any graph then* $D(H) = 2m$*, where m is the size of graph H* [2].

Example 9.1 Show that in a graph K, if $d(h) = \text{odd}, \quad \forall\ h \in S \subseteq V(K)$ then $|S| = \text{even}$.

Example 9.2 How many nodes are necessary to draw graph with $|E(P)| = 30$ and $d(w_i) = 3, \quad \forall\, w_i \in V(P)$.

Solution We have,

$$D(P) = \sum_{\forall u \in V(P)} d(u) = 2m = 2(30) = 60$$
$$\Rightarrow\ 3 + 3 + \cdots n\ times = 60$$
$$\Rightarrow 3n = 60$$
$$\Rightarrow n = 20$$

Example 9.4 Show that the simple graph H of order 20 and size 200 does not exist.

Solution The graph H must have at most $\frac{n(n-1)}{2} = \frac{20(20-1)}{2} = 190$ edges.

Example 9.5 Show that every cubic graph has even number of vertices.

Solution By the Handshaking lemma,

$$D(P) = \sum_{\forall u \in V(P)} d(u) = 2m$$
$$\Rightarrow\ 3 + 3 + \cdots n \text{ times} = \text{even}$$
$$\Rightarrow 3n = \text{even}$$
$$\Rightarrow n = \text{even}$$

9.3 Types of Graphs: Experiential Learning

(A) **Brainstorming Experiment-II**

In Experiment-I, let us explore different possible roads among cities.

(B) **Inferences-II**

The special possible inferences are as follows.

i. Every city may be joined with the same number of roads (Regular Graph).
ii. There may exist a separate road, between every pair of cities (Complete Graph).
iii. There might be the road partitions in the cities, i.e., there may not be any road among the first group of cities P, Q, R. But, there are roads that join the cities of the first group to the cities of the second group. In the second group of cities S and T, there is no any road that joins S to T (Bipartite Graph) [1, 2].

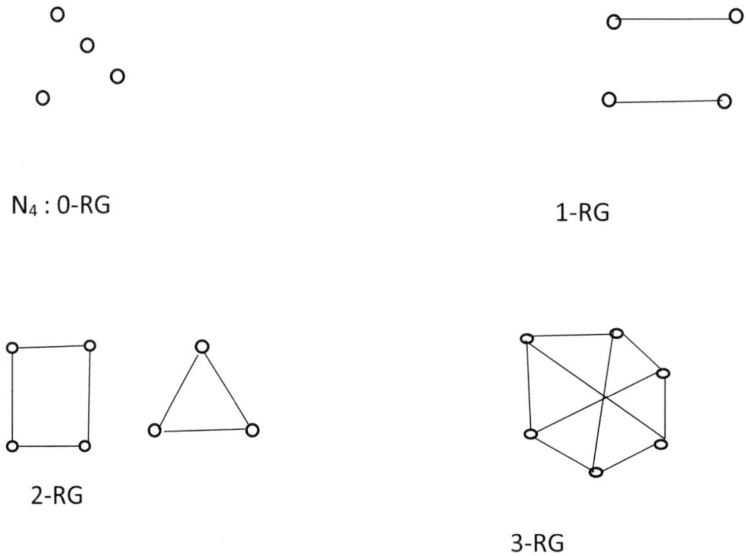

Fig. 9.2 RGraphs

(C) **Mathematical Insights-I**

1. **Regular Graph**

A graph H is called a regular graph (RG) iff $d(h) = r, \ \forall \ h \in V(H)$, where r is a constant. A RG of degree r is known as r-RG. The zero-RG is called as null graph. The cubic graph is the 3-RG.

Consider the following examples of regular graphs depicted in Fig. 9.2.

2. **Complete Graph**

A graph H is called a complete graph if every pair of vertices in H are adjacent. K_n is the complete graph with n vertices. The complete graph K_n has exactly $\frac{n(n-1)}{2}$ edges. The K_n is $(n-1)-$ RG. Let us consider the following examples of complete graphs given in Fig. 9.3.

3. **Bipartite Graph**

A graph $H(V, E)$ is said to be bipartite graph if $V = P \cup Q$ and $P \cap Q = \varphi$ such that $\forall \ e \in E(H), \ e = \{p_i, q_j\}$ such that $p_i \in P$ and $q_j \in Q$.

The bipartite graph with $|P| = r$ & $|Q| = s$ is denoted by $B_{r,s}$ or $B(r, s)$. Figure 9.4 shows the examples of bipartite graphs.

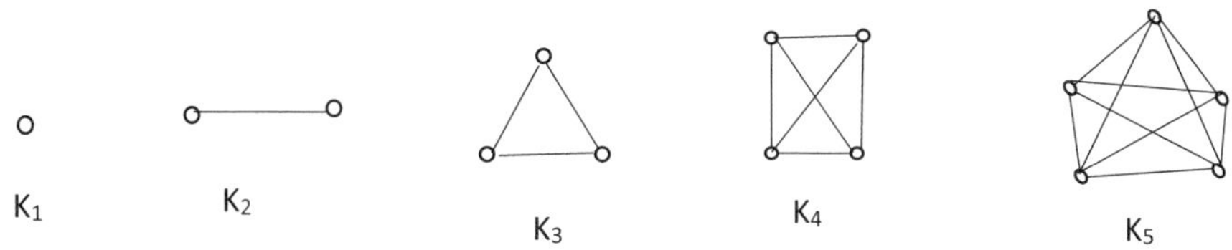

Fig. 9.3 Complete graphs

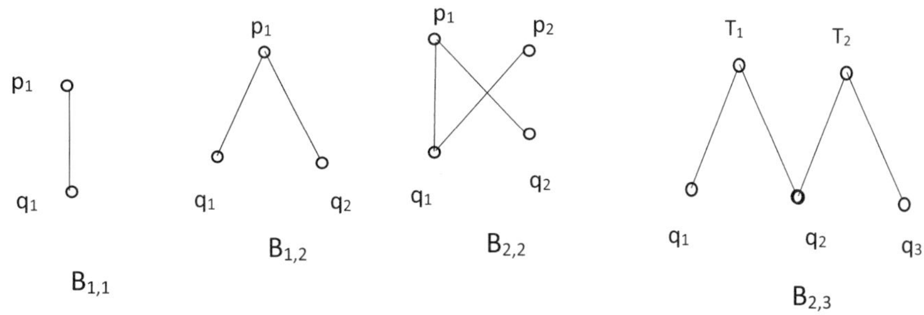

Fig. 9.4 Bipartite graphs

4. **Complete Bipartite Graph**

A graph $H(V, E)$ is said to be complete bipartite graph (CBG) if $V = P \cup Q$ and $P \cap Q = \varphi$ such that each vertex of P is adjacent with every vertex of Q. The CBG with $|P| = r$ & $|Q| = s$ is denoted by $K_{r,s}$ or $K(r, s)$.

And $K_{1,s}$ = Star graph.

Notes:

i. The complete graphs $K2$ is $K_{1,1}$ (complete bipartite graph 1 and 1 vertices).

Consider the examples of graphs depicted in Fig. 9.5.

5. **Cycle Graph**

A graph $H(V, E)$ is said to be a circuit or cycle graph with vertex set $V(H) = \{h_1, h_2, h_3, \cdots, h_n\}$ if the vertex h_r is adjacent with the vertices h_{r+1} and h_{r-1} for $r = 2, 3, \ldots, n - 1$ and the vertices h_1 and h_n are adjacent. It is denoted by C_n.

The graph $C3$ is the only graph which complete as well as cycle graph. Every cycle graph C_n is 2-regular graph.

Notes:

i. The graph $C4$ is the only cycle graph which is complete bipartite graph.
ii. The graph $C3$ is complete tripartite graph.
iii. Every cycle graph is either bipartite graph or tripartite graph.

Consider the following examples of complete n-partite graphs shown in Fig. 9.6.

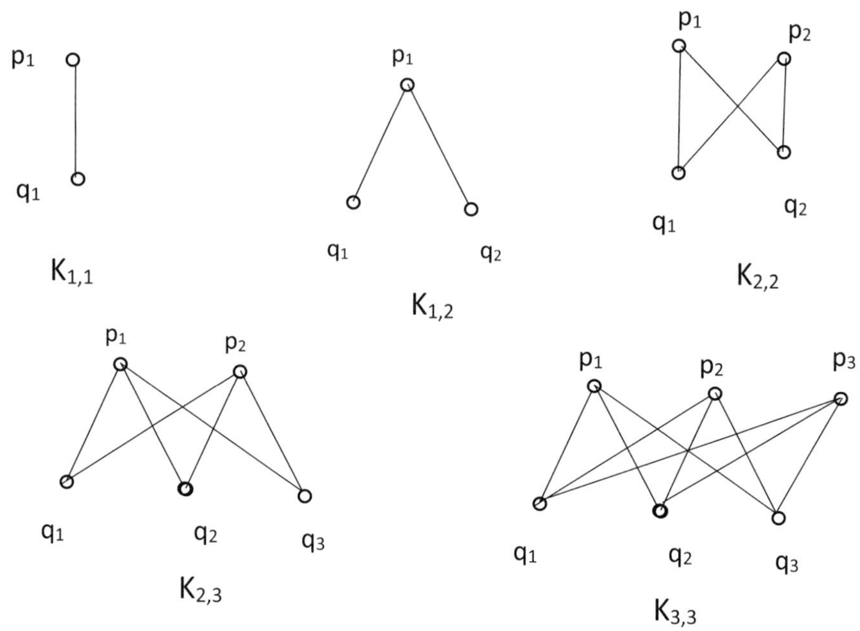

Fig. 9.5 Complete bipartite graphs

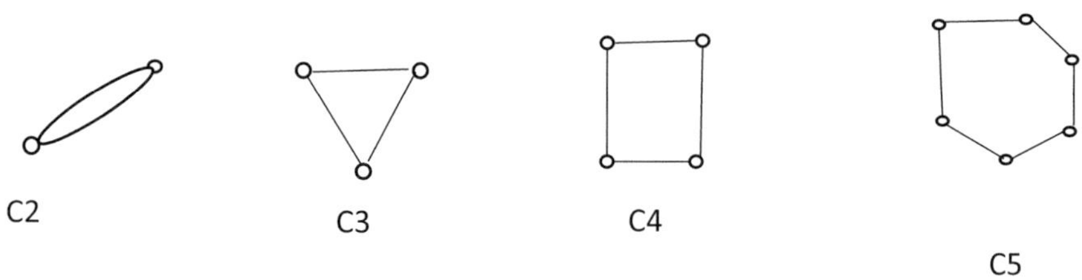

Fig. 9.6 Cycle graphs

6. **New Graphs from Old Graphs**

In this topic, we develop some new graphs from the given graphs (old graphs).

i. **Subgraphs**: A graph S is said to be a subgraph of a graph H iff $V(S) \subseteq V(H)$ and $E(S) \subseteq E(H)$ [2].

The graphs shown in Fig. 9.7 are the examples of graph and it's subgraphs. Graphs $H1$, $H2$, $H3$, $H4$, and $H5$ are subgraphs of the graph H.

i. **Edge Disjoint Subgraphs**: Two subgraphs P and M of a graph H are edge disjoint subgraphs iff $E(P) \cap E(M) = \varphi$. It may have common vertices. Subgraphs H_2 and H_3 are edge disjoint subgraphs. Any null subgraph and

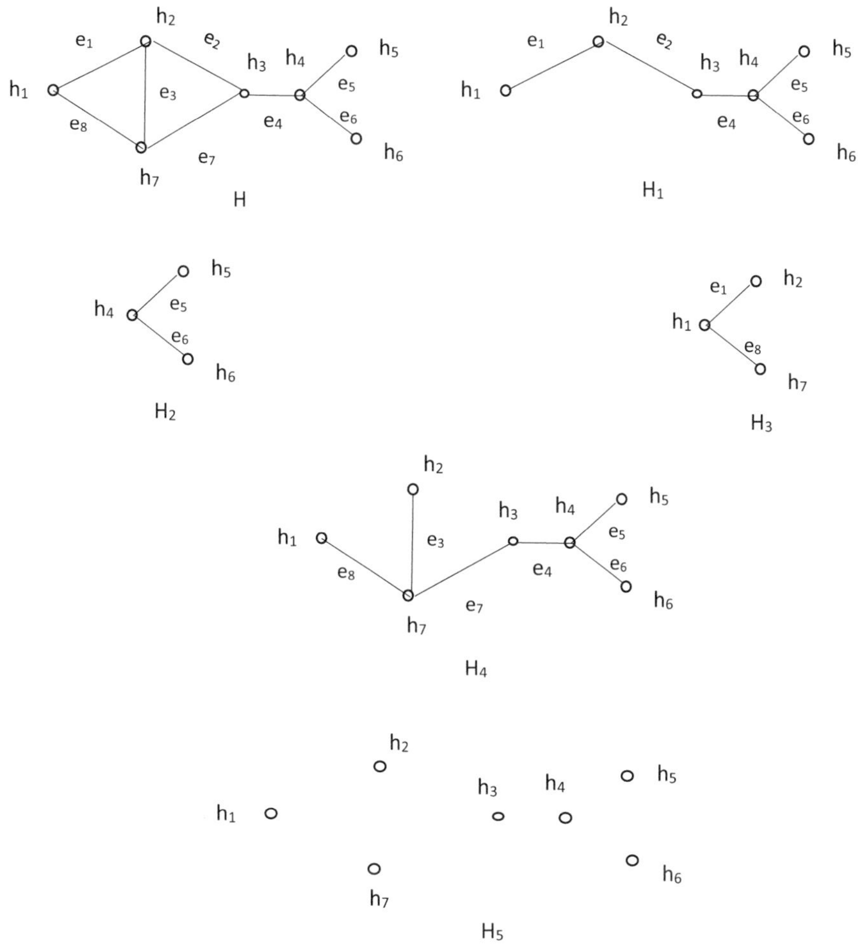

Fig. 9.7 Graph and subgraphs

subgraph of the graph are always edge disjoint subgraphs. The subgraphs H_5 and H_3, H_5 and H_1, H_5 and H_2, H_5 and H_4 are edge disjoint subgraphs.

ii. **Vertex Disjoint Subgraphs**: Two subgraphs P and M of a graph H are vertex disjoint subgraphs iff $V(P) \cap V(M) = \varphi$. All vertex disjoint subgraphs are also edge disjoint subgraphs. Subgraphs H_2 and H_3 are vertex disjoint subgraphs, but H_1 and H_3 are not vertex disjoint subgraphs. All vertex disjoint subgraphs are edge disjoint subgraphs, but the converse is not true. Subgraphs H_1 and H_5 are edge disjoint but not vertex disjoint subgraphs.

iii. **Spanning Subgraph**: A subgraph P of a graph H is a spanning subgraph of a graph H iff $V(P) = V(H)$. Figure 9.8 shows the graph and its spanning subgraphs. The subgraphs H_1, H_2, and H_3 are spanning subgraphs of the graph

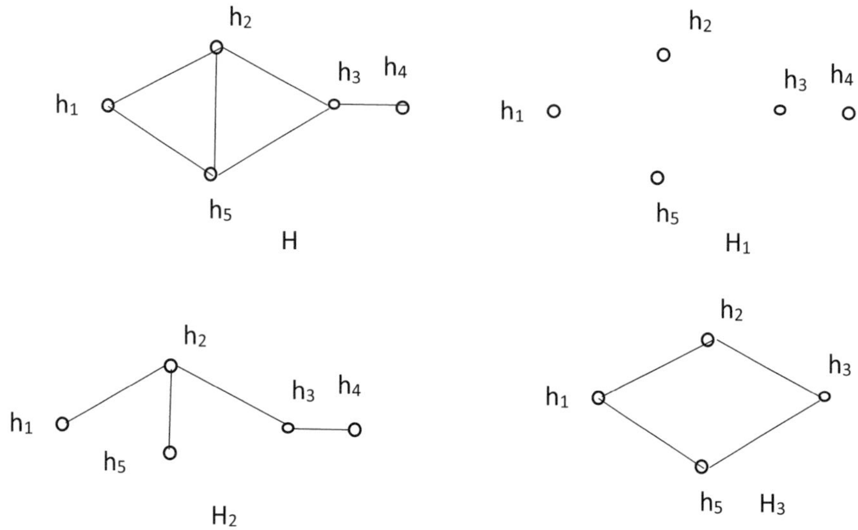

Fig. 9.8 Spanning subgraphs

H. Two spanning subgraphs may or may not be edge disjoint subgraphs. Furthermore, two spanning subgraphs never be vertex disjoint subgraphs. The subgraphs H_1 and H_2 are edge disjoint but not vertex disjoint subgraphs.

iv. **Complement of a Graph**: A graph P is a complement of a graph M iff $V(P) = V(M)$ and $E(P) = \{e = \{s, t\}/e = \{s, t\} \notin E(M)\ \}$. It is denoted by $\overline{P}$.

Properties:

1. For any graph, $\overline{(\overline{P})} = P$.
2. $\overline{K_n} = N_n$ and $\overline{N_n} = K_n$.
3. The complement of the regular graph is regular.
4. The complement of a graph with respect to the edges does not exist.

Figure 9.9 depicts the graph and the corresponding complements.

9.4 Graph Operations: Experiential Learning

In this section, we define the graph operations analogues to the operations in the set theory.

1. **Intersection of Two Graphs**

The intersection graph H of two graphs P and Q, $H = P \cap Q$ is a graph such that $V(H) = V(P) \cap V(Q)$ and $E(H) = E(P) \cap E(Q)$. It is denoted by $H = P \cap Q$.

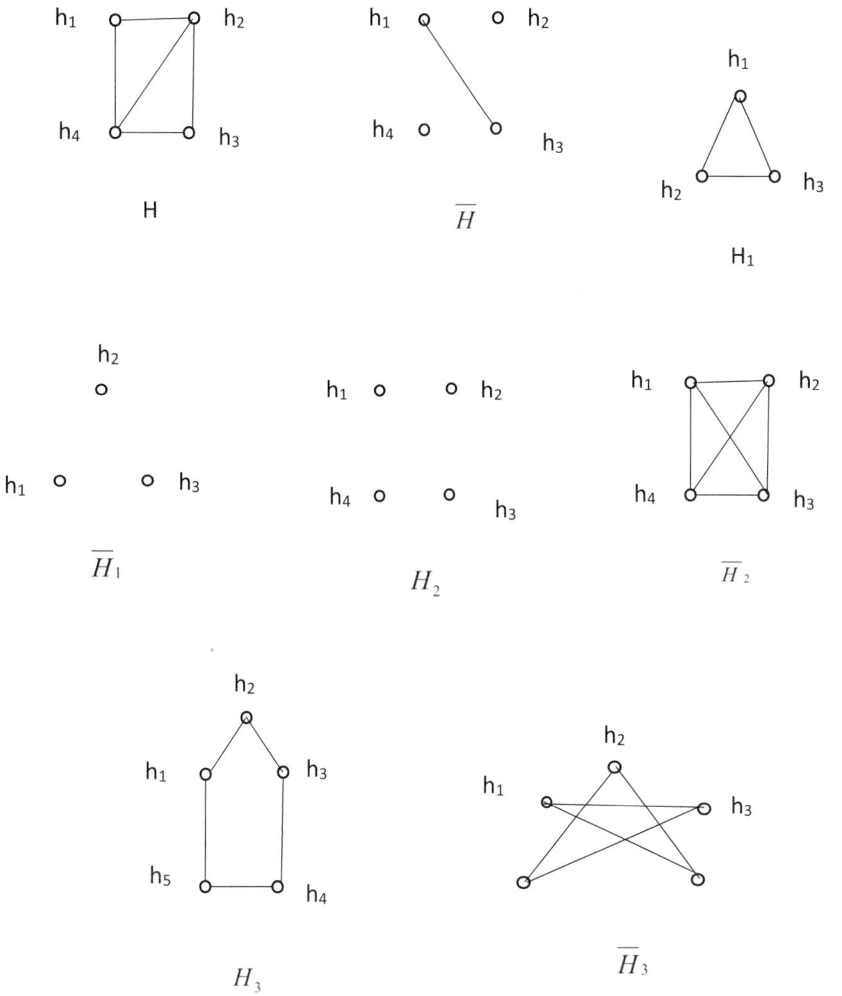

Fig. 9.9 Graphs and it's complements

Let us consider Fig. 9.10 having graphs and their intersection graph.

2. **Union of Two Graphs**: The union graph H of two graphs P and M is a graph such that $V(H) = V(P) \cup V(M)$ and $E(H) = E(P) \cup E(M)$. It is denoted by $H = P \cup M$ Fig. 9.11.

3. **The Ring Sum of Two Graphs**: The ring sum of two graphs P and M is denoted by $H = P \oplus M$ such that $V(H) = V(P) \cup V(M)$ and Fig. 9.12

$$E(H) = \{e/e \in E(P) \text{ or } E(M) \text{ but } e \notin E(P) \cap E(M)\}$$

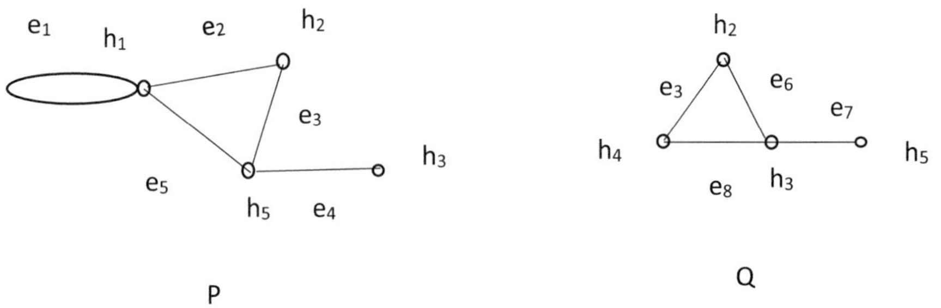

$V(P) = \{ h_1 ,\ h_2 , h_3 , h_4 \}$, $E\ (P) = \ = \{ e_1 ,\ e_2 , e_3 , e_4 , e_5 \}$
$V(Q) = \{ h_2 , h_3 , h_4 , h_5 \}$, $E\ (Q) = \ = \{ e_3 ,\ e_6 , e_7 , e_8 \}$
$V(P \cap Q) = \{ h_2 , h_3 , h_4 \}$, $E\ (P \cap Q) = \ = \{ e_3 \}$

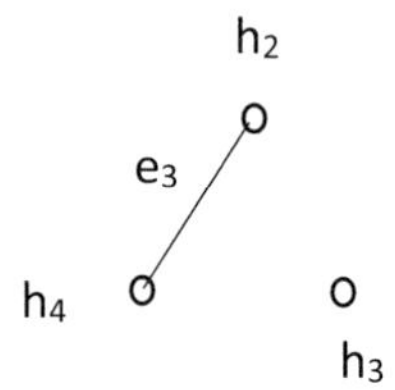

$H = P \cap Q$

Fig. 9.10 Intersection of graphs

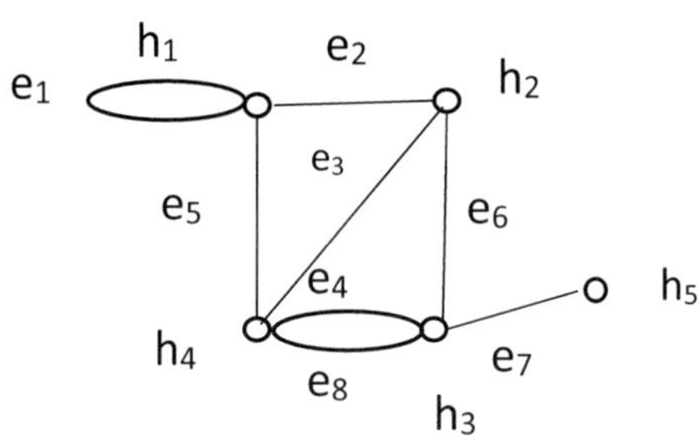

$$H = P \cup M$$

Fig. 9.11 Union of graphs

Fig. 9.12 Ring sum of gra

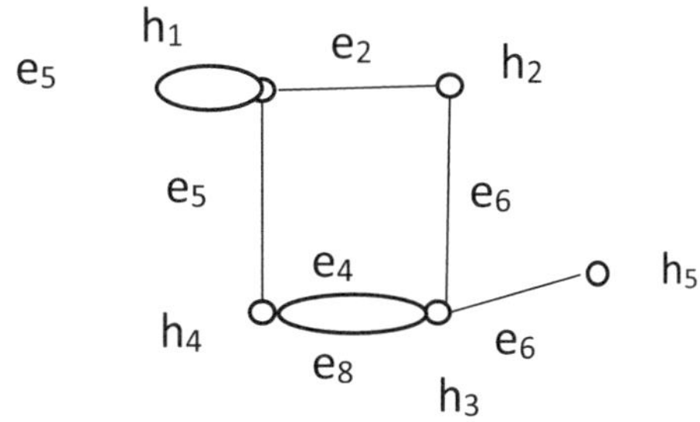

$$H = P \oplus Q$$

$$= \{E(P) \cup E(M)\} - \{E(P) \cap E(M)\}$$

4. **Sum of Two Graphs**: The sum of two disjoint graphs P and Q is denoted by $H = P + Q$ such that $V(H) = V(P) \cup V(Q)$ and

$$E(H) = \{e = \{x, y\}/e \in [E(P) \cup E(Q)] \text{ or } x \in E(P) \\ \& y \in E(Q), \forall\, x \in \text{V(P) and } \forall\, y \in \text{V(Q)}\}.$$

If the graphs $P(n_1, m_1)$ and $Q(n_2, m_2)$ then $P+Q = H(n_1+n_2,\ m_1+m_2+n_1*n_2)$.
The sum $N_n + N_m$ of null graphs is $K_{m,n}$ Fig. 9.13.
The ring sum of two regular graphs is again a regular graph.

5. **Product of Two Graphs**

The product of two graphs P and M is denoted by $H = P \times M$ such that $V(H) = V(P) \times V(M)$ and

$$E(H) = \left[\{\{p_1, m_1\}, \{p_2, m_2\}\}/\ [p_1 = p_2 \text{ and } \{m_1, m_2\} \\ \in E(M)] \text{ or } [m_1 = m_2 \text{ and } \{p_1, p_2\} \in E(P)]\right].$$

That is, $\{p_1, m_1\}$ and $\{p_2, m_2\}$ are adjacent iff

1. $p_1 = p_2$ and m_1 is adjacent to m_2 in M **or**
2. $m_1 = m_2$ and p_1 is adjacent to p_2 in P.

The product of two graphs P and M is depicted in Fig. 9.14. The product of K_2 with K_2 is C_4.

6. **Decomposition**: A graph H is said to be decomposed into subgraphs P and M iff $H = P \cup M$ and $P \cap M$ = Null graph.

The decomposition of the graph H is depicted in Fig. 9.15

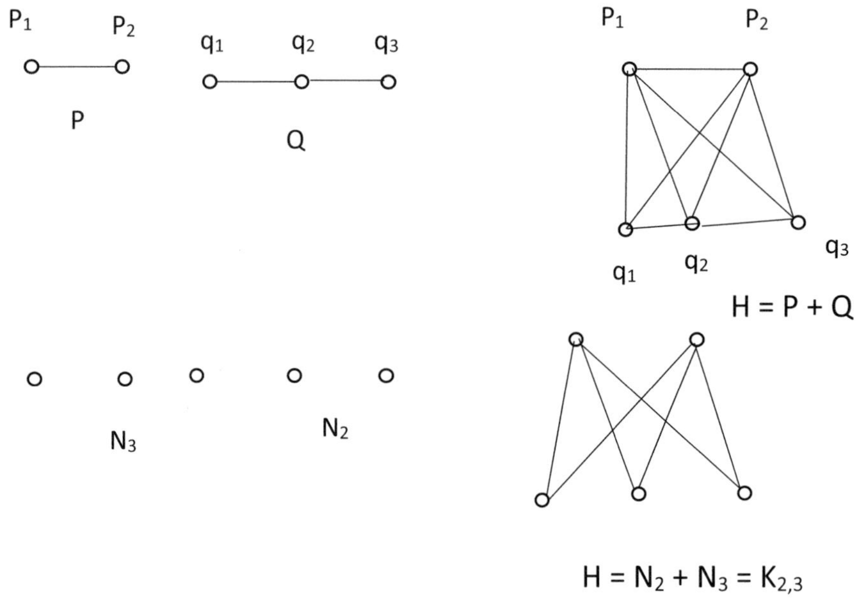

Fig. 9.13 Sum of graphs

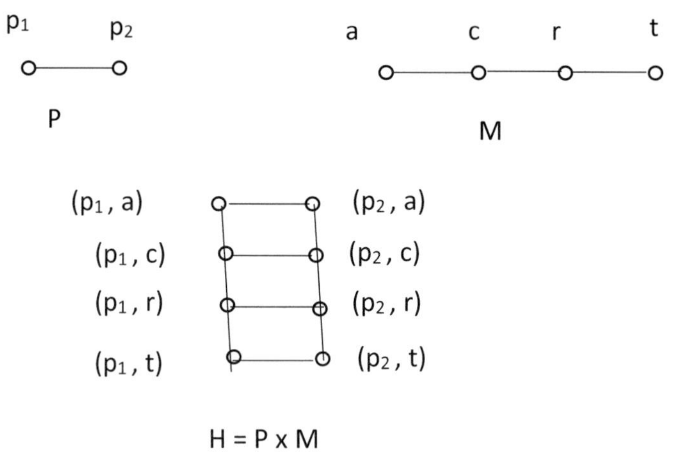

Fig. 9.14 Product of two graphs

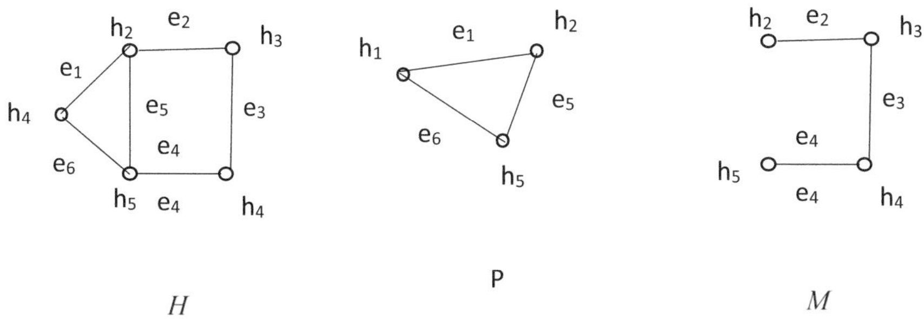

Fig. 9.15 Decomposition of graph

9.5 Isomorphism of Graphs: Experiential Learning

In real life, there are many objects which look like equal or same but they are different. Such concept is also expressed in graph theory. Two graphs are equal with respect to graph attributes. These similarities are attempted in graph theory by defining the various types of graph isomorphisms [1, 2].

Definition Two graphs H and M are said to be isomorphic graphs if

1. $\exists$ a bijective function $\varphi : V(H) \rightarrow V(M)$.
2. $\exists$ a bijective function $\psi : E(H) \rightarrow E(M)$ such that $\{p, q\} \in E(H)$ iff $\{\varphi(p), \varphi(q)\} \in E(M)$. That is, the function ψ preserves the adjacency and non-adjacency of vertices.

The functions φ and ψ are called an isomorphism of graphs H and M. It is denoted by $H \cong M$.

Suppose $H \cong M$ then

i. $|V(H)| = |V(M)|$.
ii. $|E(H)| = |E(M)|$.
iii. H and M have an equal number of vertices of degree r.
iv. If $d(h) = r$ in H then $d[\varphi(h)] = r$ in M.
v. If $\{p, q\} \in E(H)$ then $\{\varphi(p), \varphi(q)\} \in E(M)$.

If anyone of the above properties fail, then H and M are not isomorphic.

The necessary conditions for two graphs H and M to be isomorphic are

(a) $|V(H)| = |V(M)|$.
(b) $|E(H)| = |E(M)|$.
(c) H and M have an equal number of vertices of degree r.

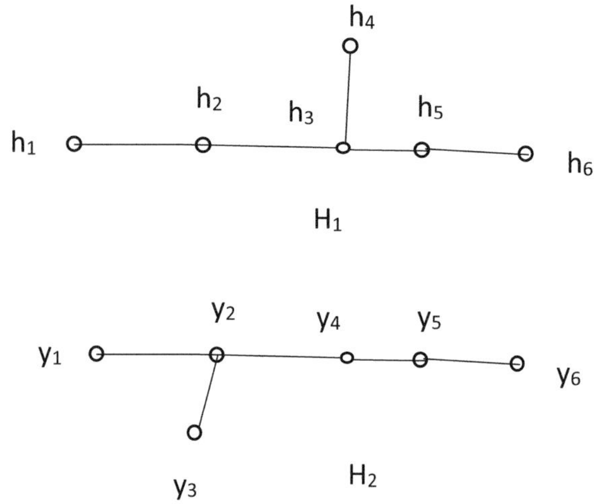

Fig. 9.16 Non-isomorphic graphs

These conditions are not sufficient. The following graphs shown in Fig. 9.16, satisfies the above conditions, but these graphs are not isomorphic as the degree of the vertex h_3 is 3 and adjacent with one vertex of degree 1 in H_1. The image of h_3 must be a vertex of degree 3, i.e., y_2 in H_2. But, the vertex y_2 is adjacent with two vertices of degree 1. Thus, the graphs $H1$ and $H2$ do not preserve the adjacency of vertices. Hence, graphs $H1$ and $H2$ are not isomorphic.

Note: The graphs $K_{m,n}$ and $K_{n,m}$ are isomorphic.

9.6 Connectivity of Graphs: Experiential Learning

Walk: Let H be any graph having $V(H) = \{h_1, h_2, h_3, \ldots, h_n\}$ and $E(H) = \{e_1, e_2, e_3, \ldots, e_m\}$. An alternating sequence of vertices and edges in H, say

$S : h_1 - e_1 - h_2 - e_2 - h_3 - e_3 - \cdots - h_r - e_r - h_{r+1}$ such that $e_i = \{h_i, h_{i+1}\}$ for $i = 1, 2, 3, \ldots, r$, is called a walk. The edge in every walk, is incident at the vertices immediately preceding it and following it [2].

The end vertices in the walk, h_1 and h_{r+1} are called as terminal vertices and other vertices are known as the interior vertices. In walk, vertices and edges may be repeated any number of times. There are many walk between every distinct pair of vertices. Depending upon the nature of the terminal vertices, there are two types of walks.

A walk in which first vertex = last vertex, is known as the closed walk. Otherwise it is open walk. Figure 9.17 depicts the graph and the walks.

The following are the examples of walks in the graph H.

(i) W_1: $h_1 - e_1 - h_2 - e_2 - h_3 - e_{10} - h_1 - e_{10} - h_3 - e_3 - h_4$
(ii) W_2: $h_1 - e_1$ - $h_2 - e_6 - h_5 - e_4 - h_4$

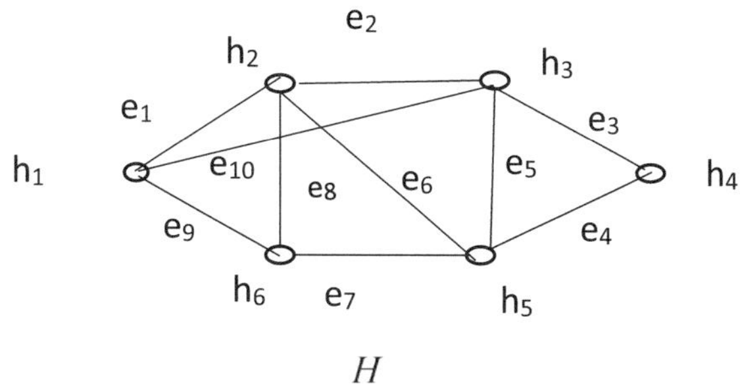

Fig. 9.17 Walk graphs

(iii) W_3: $h_1 - e_1 - h_2 - e_8 - h_6 - e_9 - h_1 - e_{10} - h_3$
(iv) W_4: $h_1 - e_{10} - h_3 - e_8 - h_4$
(v) W_5: $h_1 - e_1 - e_8 - h_6 - e_9 - h_1$

Trail: A walk of a graph H having all distinct edges, is called a Trail. The vertices may be repeated any number of types in the trail. W_1 is not a trail but W_2, W_3, W_4 are trails and W_5 is a closed trail.

Path: A walk of a graph H having all distinct vertices, is called a path. Every path P is a trail but the converse is not true. In above graph, W_1 and W_3 are not paths but W_2 and W_4 are paths.

Cycle or Circuit: A closed path is known as a cycle or circuit. W_5 is a closed path.

Connected Graph

A graph H is called a connected graph iff ∃ a walk or trail or path between every pair of vertices of H. Not connected means disconnected graph. The disconnected graph involves two or more connected parts known as the blocks or components.

In Fig. 9.18, graphs $H1$ and $K_{2,3}$ are connected graphs, whereas the graph $H2$ is regular but not connected.

Connectivity of a Graph: The connectivity of graph is associated with the number of edges or vertices whose elimination from the graph provides more components or blocks. Depending upon two attributes of the graph (Edges and Vertices), there

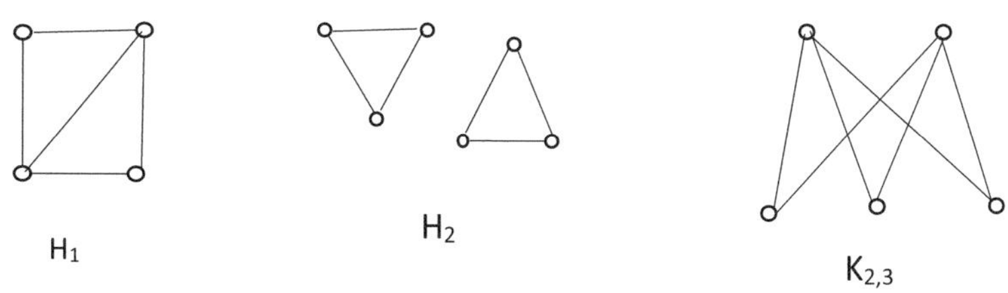

Fig. 9.18 Connected and disconnected graphs

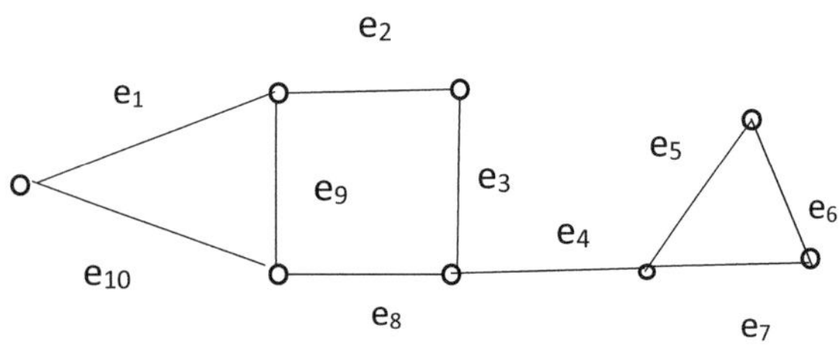

Fig. 9.19 Edge connectivity of graphs

are two types of the connectivity of graphs, known as the edge connectivity and the vertex connectivity [1].

1. **Edge Connectivity**: A set of edges D whose deletion from the connected graph H, disconnects the graph, is called as the disconnecting set. A disconnecting set D is called a minimal disconnecting if no proper subset of the set D is the disconnecting set. The minimal disconnecting set is known as the cutset of the graph. The number of edges in the smallest cutset D (Cardinality of D is minimum), is called as the edge connectivity of the graph. Notation: $\lambda(H)$.

If a cutset of a graph H contains only one edge, then that edge is called as the Isthmus or Bridge.

Consider an example of $\lambda(H)$ depicted in the Fig. 9.19.

Let us consider the following edge sets.

Sr. No.	Edge set	Remark
1	$D_1 = \{e_1, e_2, e_9\}$	Minimal disconnecting set
2	$D_2 = \{e_1, e_{10}, e_9\}$	Disconnecting set but not minimal as it's subset $\{e_1, e_{10}\}$ is the disconnecting set
3	$D_3 = \{e_4\}$	Minimal disconnecting set and e_4 is an isthmus or bridge
4	$D_4 = \{e_2, e_3\}$	Minimal disconnecting set
5	$D_5 = \{e_1, e_8, e_9\}$	Minimal disconnecting set
6	$D_6 = \{e_2, e_8, e_3\}$	Disconnecting set but not minimal as it's subset $\{e_2, e_8\}$ is the disconnecting set
7	$D_6 = \{e_2, e_1, e_6\}$	Not disconnecting set

In this graph H, the smallest cutset is $D_3 = \{e_4\}$. Therefore, $\lambda(H) = 1$.

Note

- The graph H has an isthmus iff $\lambda(H) = 1$.
- **Vertex Connectivity**

The minimum number of vertices/nodes whose deduction or deletion from the connected graph H, disconnects the graph or gives K_1, is called as the vertex connectivity [notation: $k\ (H)$] of the graph. A graph is said to be r-connected if it's $k\ (H)$ is r.

A vertex $h \in V(H)$ is cut vertex iff $V(H)$—$\{h\}$ is a disconnected graph or K_1.

Consider Fig. 9.20 of a connected graph.

Let us consider the following vertex sets.

Sr. No.	Vertex set	Remark
1	$D_1 = \{h_2, h_7\}$	Disconnecting set
2	$D_2 = \{h_1, h_2, h_7\}$	Not disconnecting set
3	$D_3 = \{h_3\}$	Disconnecting set and h_3 is a cut vertex
4	$D_4 = \{h_2, h_3, h_4, h_5, h_6\}$	Not disconnecting set

In this graph H, the smallest cutset is $D_3 = \{h_3\}$. Therefore, $k(H) = 1$.

Notes:

- If $k(H) = 1$ then H is called as separable graph. The graph H given in Fig. 9.20 is separable graph.
- The $k(K_n) = n - 1$, $k(K_{n,m}) = \min\{m, n\}$.
- The vertex connectivity of C_n for $n > 2$, is 2.

Theorem 9.1 *In a connected graph H,* $\lambda(H) \leq d_{\min}$*, where* $d_{\min} = \min\{d(h_i)/\ \forall\ h_i \in V(H)\}$.

Proof Let h be a vertex in the graph h such that $d(h) = d_{\min}$. The vertex h can be separated by deleting all edges which are incident at h, that is, $d_{\min}$ number of edges. Hence the proof.

Theorem 9.2 *In a connected graph H,* $k(H) \leq \lambda(H)$.

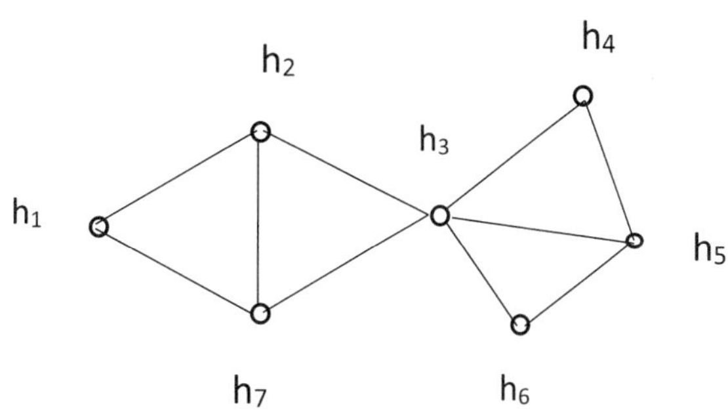

Fig. 9.20 Vertex connectivity of graphs

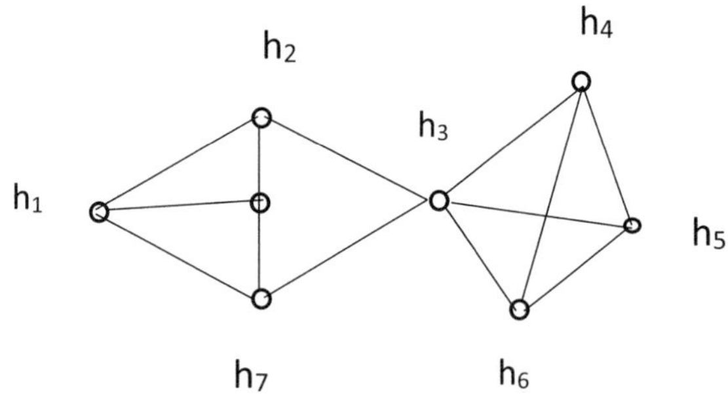

Fig. 9.21 Vertex connectivity and edge connectivity

Note: In the complete graph K_n, $k(K_n) = \lambda(K_n) = d_{\min} = n - 1$.

In the following graph $k(H) = 1,\ \lambda(H) = 2,\ d_{\min} = 3\ \Rightarrow k(K_n) < \lambda(K_n) < d_{\min}$ Fig. 9.21.

9.7 Eulerian and Hamiltonian Graphs: Experiential Learning

(A) **Brainstorming Experiment**

In the present era, the roles and responsibility of a postman are different. The postman should distribute the letters of a particular village/region/area and collect the money daily from particular members for fixed deposit scheme.

(B) **Inferences**

There are two different observations given as follows.

1. The postman should visit all routes/roads of a particular region. (Eulerian Graph family)
2. The postman should visit particular members for collecting money. (Hamiltonian Graph)

(C) **Mathematical Insights**

Definitions A trail T in a connected graph H is said to be an Eulerian Trail (ET) if $e \in T$, $\forall e \in E(H)$ without repetition. A circuit of H is called an Eulerian circuit or cycle (EC) if it is an ET and

$$\text{Starting node} = \text{ending node}$$

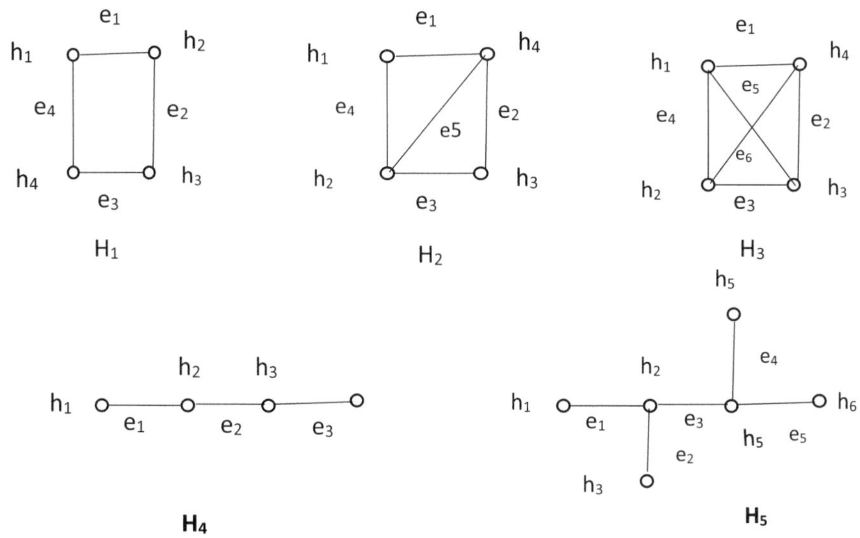

Fig. 9.22 Eulerian and non-Eulerian graphs

A graph H is said to be an Eulerian Graph (EG) if it has an Eulerian circuit. If the graph is Eulerian then the Eulerian trail, Eulerian circuit and the graph must have the same number of edges. In an Eulerian circuit, the vertices may be repeated any number of times [1].

Consider the following examples depicted in Fig. 9.22.

The graph $H1$ is Eulerian graph and the Eulerian cycle is $h_1 - e_1 - h_2 - e_2 - h_3 - e_3 - h_4 - e_4 - h_1$. The graphs $H2$, $H3$, $H4$, and $H5$ are non-Eulerian graphs. In the graph $H2$, Eulerian trail exists but not Eulerian cycle.

Notes:

- A complete graph K_n is Eulerian iff n is an odd integer.
- A r-regular graph H is Eulerian iff H is connected and r is an even integer.
- A $K_{m,n}$ is Eulerian graph iff both m and n are even.
- Every cycle graph is Eulerian.
- The bipartite graph may or may not be Eulerian.

Theorem 9.3 *A connected graph H is Eulerian iff* $d(h_i) = \text{even}, \ \forall h_i \in V(H)$, i.e.,*each vertex of H is of even degree.*

Theorem 9.4 *A connected graph H has Eulerian trail iff odd degree vertices are either 2 or 0.*

Hamiltonian Graph

We know that the Eulerian graphs are associated with the edges and Eulerian trail. Now, let's consider the second attribute of the graph, i.e., vertices to define the Hamiltonian graphs.

A path P, in graph H, is called a Hamiltonian Path (HP), if $u \in P$, $\forall u \in V(H)$ without repetition. C is Hamiltonian circuit/cycle (HC) iff $u \in C$, $\forall u \in V(H)$ without repetition and same terminal vertices. A graph having HC, is called as a Hamiltonian graph (HG) [2].

The Hamiltonian circuit must have distinct edges.

Theorem 9.4 *Ore's Theorem.*

If H is a simple graph & n > 2 vertices and $d(h) + d(m) \geq n$ *for each* $e = \{h, m\} \notin E(H)$*, then H is a Hamiltonian graph.*

Theorem 9.5 *Dirac's Theorem.*

If H is a simple graph with n > 2 vertices and $d(h) \geq \frac{n}{2}$, $\forall\, h \in V(H)$ *then H is a Hamiltonian graph.*

Example 9.6 Show that the graph K_n $(n > 2)$ is Hamiltonian. What is the length of Hamiltonian circuit? How many circuits exist in K_n?

Solution The graph K_n has n vertices and degree of each vertex is $n - 1$. Therefore, $d(h) = n - 1 \geq \frac{n}{2}$; $\forall h \in V(K_n)$.

By Theorem 9.6, K_n has a Hamiltonian circuit and hence, it is Hamiltonian graph.

The Hamiltonian circuit contains all the vertices of K_n. Therefore, its length is *n*. But there are $\frac{n(n-1)}{2}$ edges in the graph K_n. Hence, there are $\frac{(n-1)!}{2}$ distinct Hamiltonian circuits.

Example 9.7 Give two examples of the following graphs

1. Eulerian (EG) and Hamiltonian (HG)
2. EG, but not HG
3. HG, but not EG
4. Neither EG nor HG

Solution

1. EG and HG Fig. 9.23

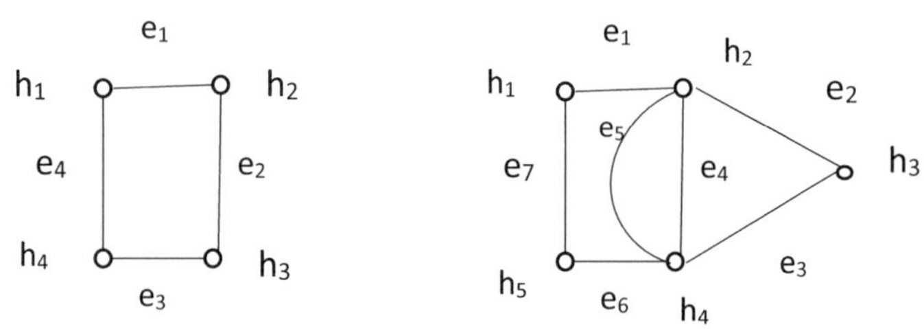

Fig. 9.23 EG and HG

1. EG, but not HG Fig. 9.24
2. HG, but not EG Fig. 9.25
3. Neither EG nor HG

Example 9.8 Draw at least seven example of connected simple non-isomorphic graphs on five vertices which are neither Eulerian nor Hamiltonian Fig. 9.27.

Solution:

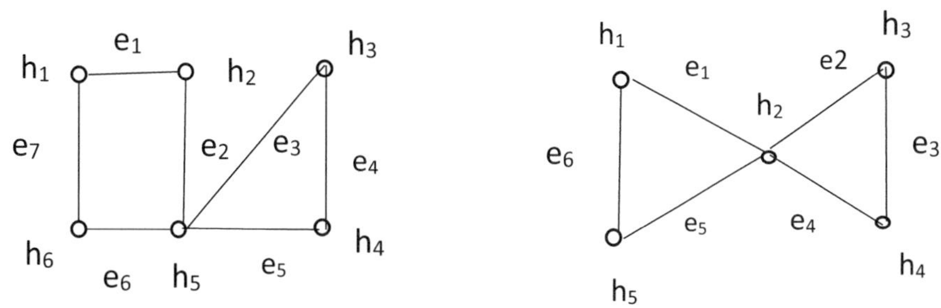

Fig. 9.24 EG but not HG

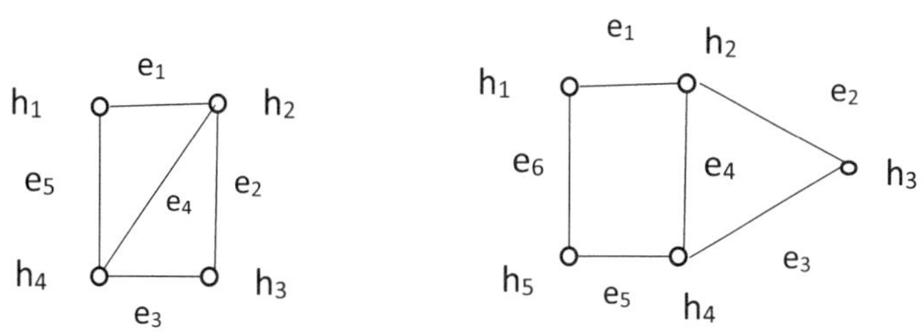

Fig. 9.25 HG but not EG

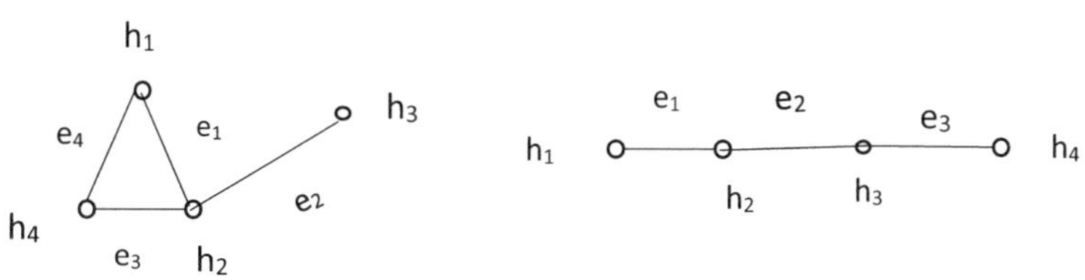

Fig. 9.26 Neither EG nor HG Fig. 9.26

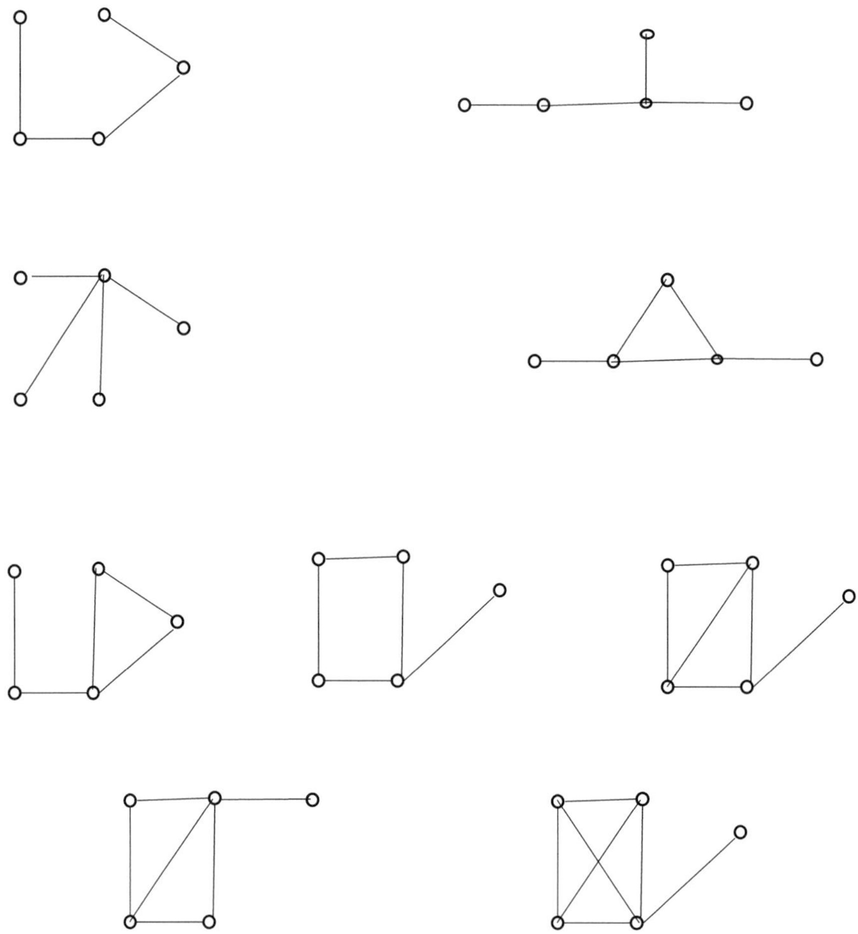

Fig. 9.27 Neither Eulerian nor Hamiltonian and non-isomorphic graphs

9.8 Weighted Graph: Experiential Learning

A graph H is said to be a weighted graph, if every edge is associated with some numeric value.

$$\text{Weight of Graph} = \sum_{\forall h \in V(H)} w\,(h)$$

Consider Fig. 9.28 of weighted graphs and their weights.

Fig. 9.28 Weighted graph

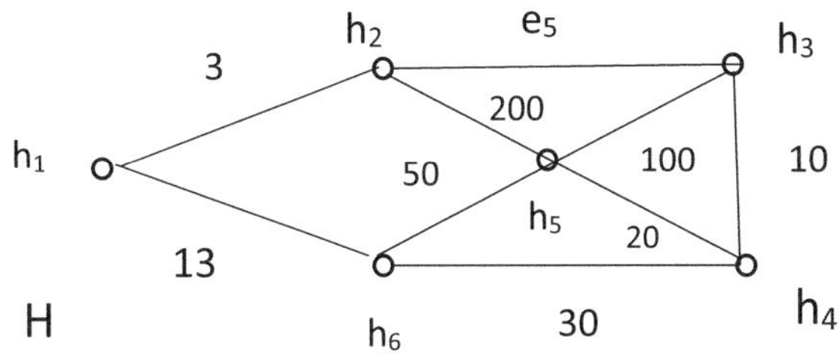

9.9 Planar Graph: Experiential Learning

This section deals with the drawing of graphs without crossing the edges on plane surface (2-dimensional) and some important results related to such kinds of graphs.

Definition A graph P is a planar graph if H can be drawn on a plane such that no edges intersect or cross in a point other than their end vertices. Otherwise, it is called as non-planar graph.

Figure 9.29 depicts the examples of planar and non-planar graphs [6].

The graphs H_1 and H_2 are planar graphs but the graph $H_3 = K_5$ is non-planar.

Regions: A planar graph H splits the plane into parts (regions). The regions are also known as faces or windows or meshes. The region is bounded by the set of edges, which forms its boundary. A region is finite, if its area is finite. Otherwise, it is

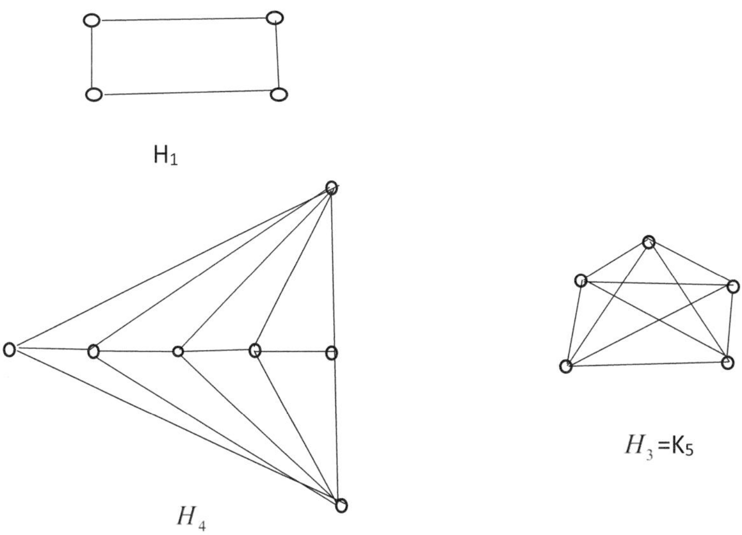

Fig. 9.29 Planar and non-planar graphs

infinite or unbounded. Every planar graph has finite regions and an infinite region. Figure 9.30 depicted the regions.

Theorem 9.6 *Euler's Formula/Theorem.*

If H is a connected planar graph then

1. $|V(H)| + |R(H)| - |E(H)| = 2$
2. $|E(H)| \leq 3\,|V(H)| - 6$
3. $|E(H)| \leq 2\,|V(H)| - 4$ *if H is triangle free graph.*

Notes:

1. *If in a graph H,* $e \leq 3v - 6$ *is not true* $(e > 3v - 6)$ *then H is not a simple connected planar graph.*
2. *If in a graph h,* $e \leq 2v - 4$ *is not true* $(e > 2v - 4)$ *then H is not a simple connected planar and triangle free graph.*

Example 9.9 Prove that K_5 (complete graph on 5 vertices) is not planar graph.

Solution The graph K_5 has 5 vertices and 10 edges. Therefore, $v = 5$ and $e = 10$. We have, $3v - 6 = 15–6 = 9$ and $e = 10$. So, $e > 3v - 6$. Thus by Corollary 1, the graph K_5 is not planar.

Example 2 Prove that $K_{3,3}$ (complete bipartite graph) is not planar graph.

Solution The graph $K_{3,3}$ has 6 vertices and 9 edges and it is triangle free graph. We have, $2v - 4 = 12–4 = 8$ and $e = 9$. So, $e > 2v - 4$. Thus by corollary 4, the graph $K_{3,3}$ is not planar.

Note: The Graph K_5 is the Kuratowski's first graph as it is smallest non-planar graph with respect to the number of vertices. The Graph $K_{3,3}$ is called the Kuratowski's second graph as it is smallest non-planar graph with respect to the number of edges Fig. 9.31.

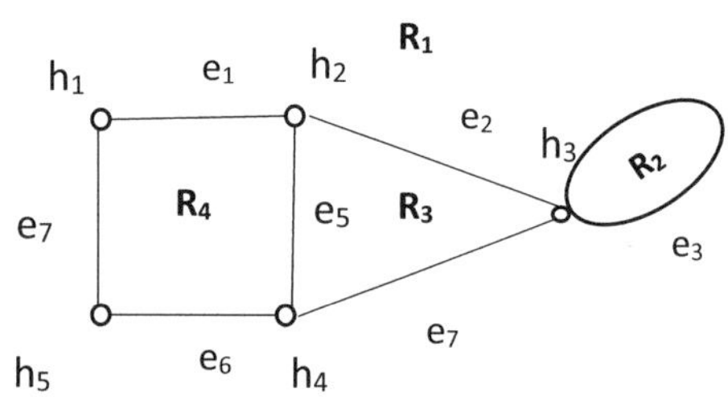

Fig. 9.30 Planar graph with regions

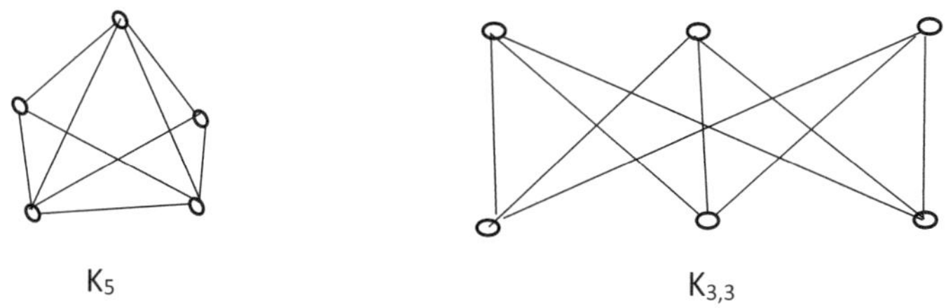

Fig. 9.31 Smallest non-planar graphs

9.10 Graph Coloring: Experiential Learning

The coloring of graph is dealing with the coloring of the graph attributes, i.e., vertices or edges or regions or vertices and edges or vertices, edges and regions according to the concerned theory.

There are following different types of coloring of graphs [7, 8].

1. Vertex coloring.
2. Edge coloring.
3. Region coloring.
4. Vertex and edge coloring (Total Coloring).
5. Vertex, edge, and region coloring.
6. π coloring.

Let us consider the coloring types one by one.

1. Vertex coloring.

Proper vertex coloring (PVC) means the adjacent vertices have different colors, i.e., $e = \{h_i, h_j\} \in E(H) \Rightarrow c(h_i) \neq c(h_j), \ \forall \ i\&j$ where $c(h_i)$ means the color of the vertex h_i.

The chromatic number $\chi(H) =$ the minimum number of colors in PVC. A graph H is said to be p-colorable if $\chi(H) \leq p$.

This type of coloring is easy, if there are less number of vertices. But for large number of vertices, this process becomes challenging.

Let us consider the following examples of the coloring of graphs depicted in Fig. 9.32.

Important results:

1. $\chi(H) = 1$ iff H is a null graph.
2. $\chi(K_n) = n$
3. $\chi(K_{m,n}) = 2$
4. For cycle graph C_n,$\chi(C_n) = 2$, if n is even and 3, if n is odd.
5. The bipartite graph is 2-colorable.
6. The graph H is n-partite iff $\chi(H) = n$.

2. **Edge Coloring**: The coloring of all edges of a graph H, such that the adjacent edges receives the different colors is called the edge coloring of graph, i.e.,

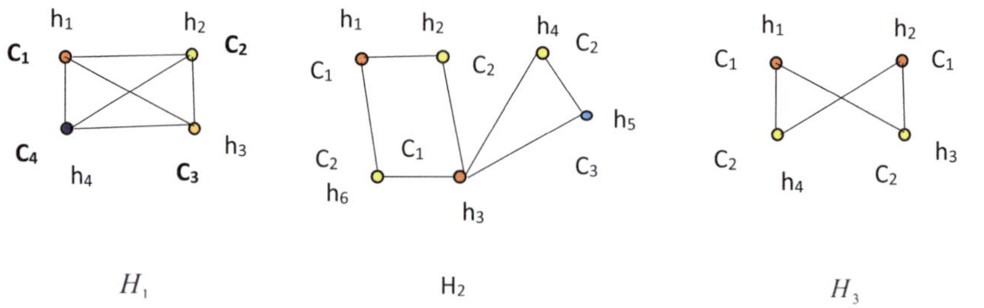

Fig. 9.32 Vertex coloring of graphs

$e_p = \{h_i, h_j\} \& e_q = \{h_j, h_k\} \in E(H) \Rightarrow c(e_p) \neq c(e_q), \ \forall \, p\&q$ and $i \neq j \neq k$ where $c(e_p)$ means the color of the edge e_p.

The edge chromatic number or chromatic Index of the graph H, is $\chi'(H) =$ minimum number of colors such that $e_p = \{h_i, h_j\} \& e_q = \{h_j, h_k\} \in E(H) \Rightarrow c(e_p) \neq c(e_q), \ \forall \, p\&q$ Fig. 9.33.

3. **Region Coloring**: Let H be a planar graph. Two regions R_1 and R_2 are adjacent if $B(R_1) \cap B(R_2) \neq \varphi$ here, $B\,(R_1) =$ set of boundary edges in R_1 [2].

   Non-adjacent regions means $B(R_1) \cap B(R_2) = \varphi$.

   The coloring of all regions of a graph H such that

   i.e., $B(R_p) \cap B(R_q) \neq \varphi \Rightarrow c(R_p) \neq c(R_q), \ \forall \, p\&q$ where $c(R_p)$ means the color of the region R_p.

   The region chromatic number is $\chi_R(H) =$ minimum number of colors such that $B(R_p) \cap B(R_q) \neq \varphi \Rightarrow c(R_p) \neq c(R_q), \ \forall \, p\&q$ Fig. 9.34.

   Important results:

   - For Complete graphs, $\chi_R(K_1) = 1, \ \chi_R(K_2) = 1, \chi_R(K_3) = 2, \chi_R(K_4) = 4,$ and

   $\chi_R(K_r) =$ not defined for $r \geq 5$ as K_r is non-planar graph.

   - For cycle graph, $\chi_R(C_n) = 2$

4. **Total Coloring**: Let H be any graph. The coloring of $V(H)$ and $E\,(H)$ such that

   i. $e = \{h_i, h_j\} \in E(H) \Rightarrow c(h_i) \neq c(h_j), \ \forall \, i\&j$

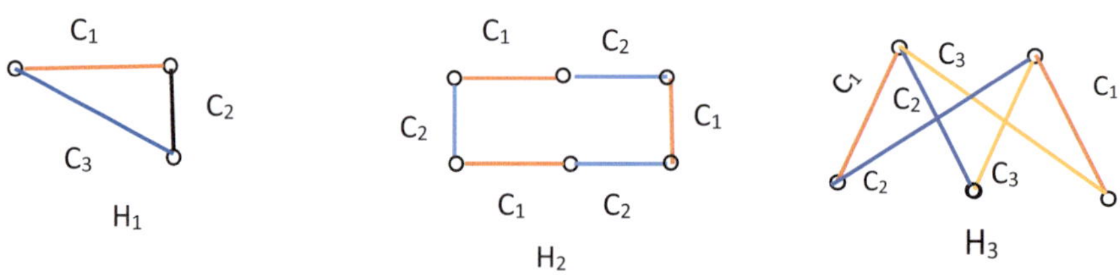

Fig. 9.33 Edge coloring of graphs

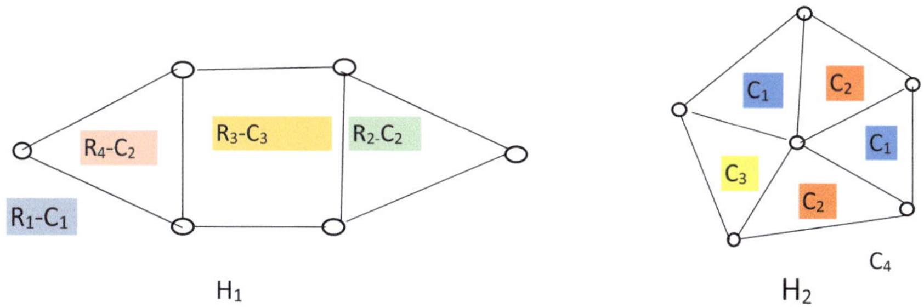

Fig. 9.34 Region coloring of graphs

ii. $e_p = \{h_i, h_j\} \& e_q = \{h_j, h_k\} \in E(H) \Rightarrow c(e_p) \neq c(e_q), \ \forall \ p\&q$ and $i \neq j \neq k$
iii. $e = \{h_i, h_j\} \in E(H) \Rightarrow c(h_i) \neq c(h_j) \neq c(e), \ \forall \ e, i\&j$

The total chromatic number of H is $\chi_T(H)$ = minimum colors satisfy above conditions.

Consider the following examples given in Fig. 9.35.

5. **Perfect Coloring**: Let H be a planar graph. The perfect chromatic number or index of a graph H is denoted by $\chi_P(H)$ and it is the minimum number of colors required to color the graph H such that

i. $e = \{h_i, h_j\} \in E(H) \Rightarrow c(h_i) \neq c(h_j), \ \forall \ i\&j$
ii. $e_p = \{h_i, h_j\} \& e_q = \{h_j, h_k\} \in E(H) \Rightarrow c(e_p) \neq c(e_q), \ \forall \ p\&q$ and $i \neq j \neq k$

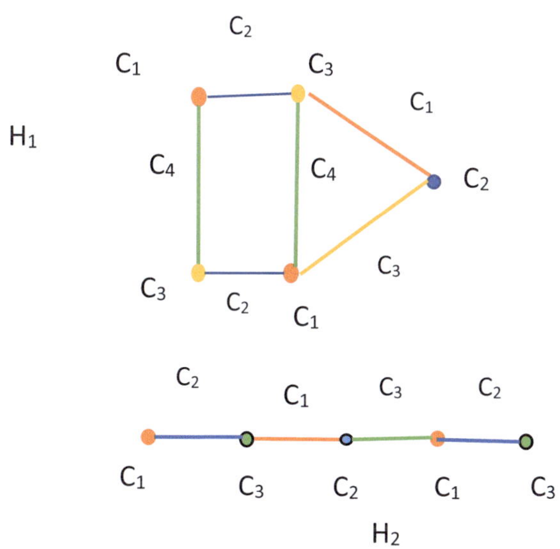

Fig. 9.35 Total coloring of graphs

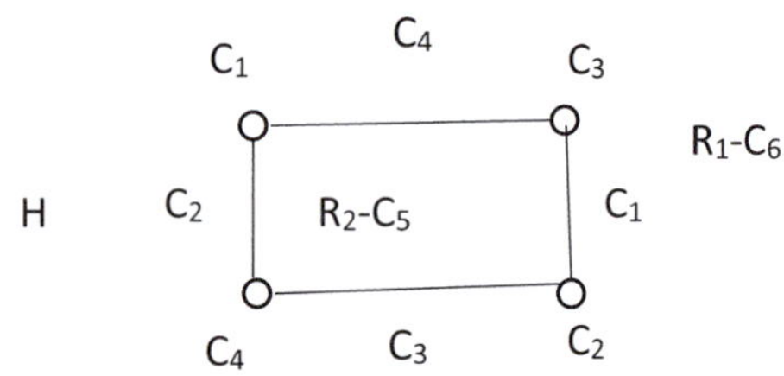

Fig. 9.36 Perfect coloring of graphs

iii. $R_p \cap R_q \neq \varphi \Rightarrow c(R_p) \neq c(R_q), \ \forall \ p\&q$

iv. The region should have different color than boundary edges and boundary vertices [8–11].

Consider the following example depicted in Fig. 9.36.

6. **Adjacent Vertex π Coloring**: Let H be a simple graph with n vertices. Let H_i be the set of vertices adjacent to the i-th vertex. Let $X = \{H_1, H_2, H_3, \ldots, H_n\}$ be the set of subset of $V(H)$. Let $P(C)$ be the power set of the color set C having s distinct colors. If there exists a function $\pi : X \rightarrow P(C)$ such that assign colors to vertices in H such that $\pi(H_p) \neq \pi(H_q)$ for all $p \neq q$.

This coloring is known as the adjacent vertex π coloring. The smallest value of s is called as the adjacent vertex π chromatic number of a graph H [12].

It is denoted by AV π CN(H) or $\chi_\pi(H)$.

Let us consider the following examples depicted in Fig. 9.37.

Now assign colors to each vertex of H as follows. $a - 1, b - 2, c - 3, d - 4, e - 3$.

As the result, $f\ (H_1) = \{1, 2, 3\}, f(H_2) = \{1, 2, 3, 4\}, f(H_4) = \{2, 3, 4\}$, and $f(H_5) = \{1, 4, 3\}$.

Therefore, the AV π CN $(H) = 4$.

Note: For above graph H, $\chi\ (H) = 3$, which is different from AV π CN (H). Therefore, chromatic number and the AV π CN, are two different concepts.

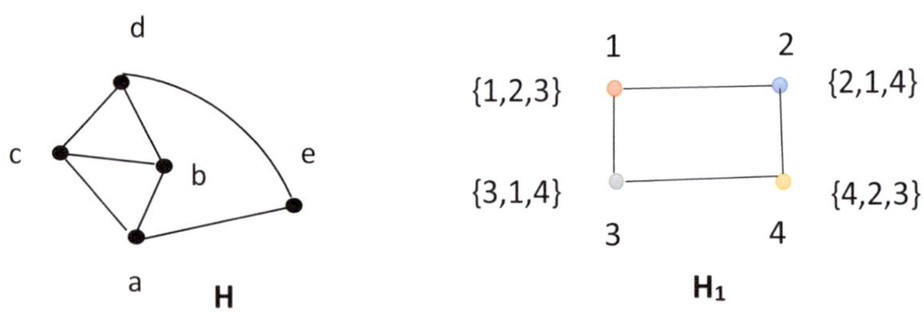

Fig. 9.37 Adjacent vertex π coloring of graphs

9.11 Representing Graphs in Computer Memory: Experiential Learning

We have seen that graphs can be represented either set theoretically or diagrammatically. But, it is difficult to represent the graphs in computer in either way. We can store the data of graphs in computer by using matrices. So, it is essential to concert graphs into matrices and vice versa. The following are main two types of matrices named as adjacency matrix and incidence matrix [13, 14].

1. **Adjacency Matrix**

Let H be any graph without parallel edges and having n vertices. The adjacency matrix of H is denoted by $A(H) = (h_{pq})_{n\times n}$ and defined as

$$\begin{aligned} h_{pq} &= 1\ ;\ \{h_p, h_q\} \in E(H) \text{ i.e. } h_p \text{ and } h_q \text{ are adjacent in } H \\ &= 0\ ;\ \{h_p, h_q\} \notin E(H) \text{ i.e. } h_p \text{ and } h_q \text{ are not adjacent in } H \end{aligned}$$

Let us consider the following examples depicted in Fig. 9.38.
The adjacency matrices of the graphs $H1$ and $H2$ are as follows.

$$A(H_1) = (h_{pq})_{4\times 4} = \begin{matrix} h_1 \\ h_2 \\ h_3 \\ h_4 \end{matrix} \begin{bmatrix} 0 & 1 & 0 & 1 \\ 1 & 0 & 0 & 1 \\ 0 & 0 & 0 & 1 \\ 1 & 1 & 1 & 0 \end{bmatrix} \qquad A(H_2) = (h_{pq})_{4\times 4} = \begin{matrix} h_1 \\ h_2 \\ h_3 \\ h_4 \end{matrix} \begin{bmatrix} 0 & 1 & 0 & 1 \\ 1 & 0 & 1 & 1 \\ 0 & 1 & 1 & 1 \\ 1 & 1 & 1 & 0 \end{bmatrix}$$

Notes:

i. $A(H)$ is a symmetric and binary matrix.
ii. In $A(H)$, $h_{pp} = 1$ if there is a loop at the p-th vertex.

2. **Incidence Matrix**

Let be any graph without loops and having n vertices and m edges. The incidence matrix of H is defined as $I(H) = (h_{pq})_{n\times m}$ and

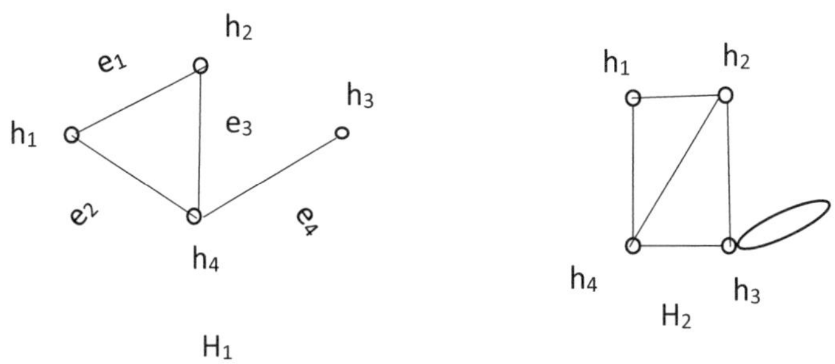

Fig. 9.38 Graphs for adjacency matrix

$$(h_{pq})_{n\times m} = 1 \; ; \; e_q = \{h_p, h_s\} \in E(H) \text{ i.e. } e_q \text{ is incident at the vertex } h_p \text{ in } H$$
$$= 0 \; ; \; e_q = \{h_p, h_s\} \notin E(H) \text{ i.e. } e_q \text{ is not incident at the vertex } h_p \text{ in } H$$

The incidence matrix of the graph $H1$ presented in Fig. 9.38 is as follows.

$$A(H_1) = (h_{pq})_{4\times 4} = \begin{array}{c} \\ h_1 \\ h_2 \\ h_3 \\ h_4 \end{array} \begin{array}{c} e_1\; e_2\; e_3\; e_4 \\ \begin{bmatrix} 1 & 1 & 0 & 0 \\ 1 & 0 & 1 & 0 \\ 0 & 0 & 0 & 1 \\ 0 & 1 & 1 & 1 \end{bmatrix} \end{array}$$

Properties of Incidence Matrix

i. It is a binary matrix.
ii. $\sum_{\forall p} h_{pq} = 2$, for all q.
iii. $\sum_{\forall q} h_{pq} = d(v_p)$.
iv. Two columns are identical iff the corresponding edges are parallel.
v. A row having all zeros is representing the isolated vertex.
vi. A row with single 1 represents a pendent vertex.

9.12 Graph Algorithms: Experiential Learning

In real life, we come across many situations or problems which we need step by step thinking and solutions to obtain more pragmatic results. The graph algorithm is associated with solving numerous difficult problems of graph theory by using the standard procedure. These algorithms help us to resolve the problems in innovative ways within a minimum time and steps. There are mainly two types of algorithms the shortest path algorithm and the traveling postman problem. Let's study these algorithms one by one [15, 16].

9.13 Shortest Path Algorithms

The name itself suggests that we have to find the shortest path among all existing paths of a graph. In this method, we express the real problem in the graph and then apply the shortest path algorithm for finding the optimum solution of it. There are mainly three algorithms used for finding the shortest path between the vertices namely Dijkstra's, Bellman-Ford, Floyd-Warshall Algorithms [17].

Dijkstra's Algorithm

(A) Dijkshtra's algorithm

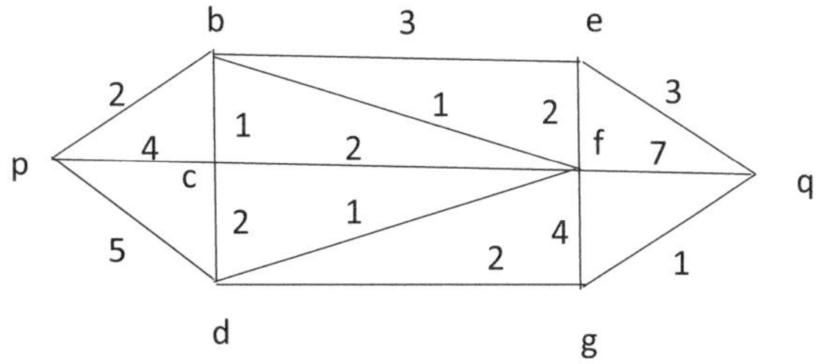

Fig. 9.39 Graph for Dijkstra's algorithm

Let $H =$ Weighted Graph & $p, q \in V(H)$ and SP= Shortest Path

Suppose, $L(x) =$ label of p which signifies the length of the shortest path. $W_{ij} =$ weight of an edge $e_{ij} = (h_i, h_j)$

Consider the following steps:

Step 1: Let S =permanent labels vertices and $T = V(H)$

Set $L(p) = 0$, and $L(x) = \infty$: $\forall\, x \in T$ and $x \neq p$

$S = \phi$ and $T = V(H)$

Step 2: Select the vertex g in $\dot{T}$ having smallest label and send g to S.

Therefore, $S = S \cup \{h\}$ and $T = T - \{h\}$.

If $g = q$ then $L(q)$= length of the minimum path from the vertex p to q and stop the procedure.

Step 3: If $g \neq q$, then revise the labels of the vertices of T. The new label is

$$L(x) = \min\{\text{old } L(x), L(g) + w(g, x)\}$$

where $w(g, x) =$ weight of edge $\{g, x\}$. If no edge linking g & x then take $w(g, x) = \infty$

Step 4: Reiteration of the steps 2 and 3 until q acquires the permanent label.

Example 9.10 Find the shortest path between p–q for the given graph H using Dijkstra's algorithm Fig. 9.39.

Solution **Step 1**: Set $S = \phi$, $T = \{p, b, c, d, e, f, g, q\}$

$$L\{p\} = 0$$

$$L\{x\} = \infty, \forall x \varepsilon T, x \neq p$$

Step 2: $h = p$, the permanent label of p is 0

$$S = \{p\}, T = \{b, c, d, e, f, g, q\}$$

$$\begin{aligned} L\{b\} &= \min\{\text{old } L(b),\ L(p) + w\,(p, b)\} \\ &= \min\{\infty, 0 + 2\} = 2 \end{aligned}$$

$$L\{c\} = \min\{\infty, 0 + 4\} = 4$$

$$L\{d\} = \min\{\infty, 0 + 5\} = 5$$

$$L\{e\} = L\{f\} = L\{g\} = L\{q\} = \infty$$

$\therefore L\{b\} = 2$ is the minimum label. The permanent label of b is 2.
Step 3: $h = b$,

$$S = \{p, b\}, T = \{c, d, e, f, g, q\}$$

$$\begin{aligned} L\{c\} &= \min\{L(c), L(b) + w\{b, c\}\} \\ &= \min\{4, 2 + 1\} = 3 \end{aligned}$$

$$L\{d\} = \min\{5, 2 + \infty\} = 5$$

$$L\{e\} = \min\{\infty, 2 + 3\} = 5$$

$$L\{f\} = \min\{\infty, 2 + 1\} = 3$$

$$L\{e\} = L\{z\} = \infty$$

$\therefore L\{c\} = L\{f\} = 3$ are the minimum labels.
Step 4: $h = c$ or f,
$h = f$, permanent label of f is 3.

$$S = \{p, b, f\}, T = \{c, d, e, g, q\}$$

$$L\{c\} = \min\{3, 3 + 2\} = 3$$

$$L\{d\} = \min\{5, 3 + 1\} = 4$$

$$L\{e\} = \min\{5, 3 + 2\} = 5$$

$$L\{g\} = \min\{\infty, 3 + 4\} = 7$$

$$L\{q\} = \min\{\infty, 3 + 7\} = 10$$

$L\{c\} = 3$ is the minimum label.
Step 5: $h = c$, the permanent label of $c = 3$

$$S = \{p,\ b,\ f,\ c\}, T\ = \{d,\ e,\ g,\ q\}$$

$$L\{d\} = \min\{4, 3 + 2\} = 4$$

$$L\{e\} = \min\{5, 3 + \infty\} = 5$$

$$L\{g\} = \min\{7, 3 + \infty\} = 7$$

$$L\{q\} = \min\{10, 3 + \infty\} = 10$$

$L\{d\} = 4$ is the minimum label.
Step 6: $v = d$, the permanent label of $d = 4$

$$S = \{p,\ b,\ f,\ c,\ d\}, T = \{e,\ g,\ q\}$$

$$L\{e\} = \min\{5, 4 + \infty\} = 5$$

$$L\{g\} = \min\{7, 4 + 2\} = 6$$

$$L\{q\} = \min\{10, 4 + \infty\} = 10$$

$L\{e\} = 5$ is the minimum label.
Step 7: $h = e$, the permanent label of e is 5

$$S = \{p,\ b,\ f,\ c,\ d,\ e\}, T = \{g,\ q\}$$

$$L\{g\} = \min\{6, 5 + \infty\} = 6$$

$$L\{q\} = \min\{10, 5 + 3\} = 8$$

$L\{g\} = 6$ is the minimum label.
Step 8: $h = g$, the permanent label of g is 6

$$S = \{p,\ b,\ f,\ c,\ d,\ e,\ g\}, T = \{q\}$$

$L\{q\} = \min\ \{8, 6 + 1\} = 7$ which is the minimum label.
Step 9: $h = q$, the permanent label of q is 7.

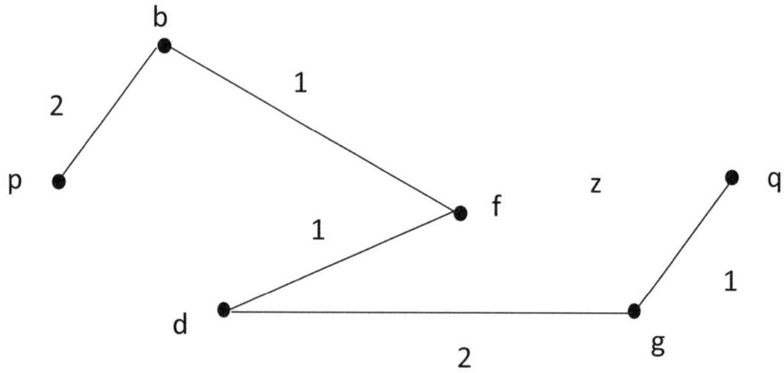

Fig. 9.40 Shortest path graph by Dijkstra's algorithm

$$S = \{p, b, f, c, d, e, g, q\}, T = \Phi$$

Hence the length of shortest path from p to q is 7.
The shortest path is $p - b - f - d - g - q$ and it is depicted in the Fig. 9.40

9.14 Case Studies and Projects

Let us consider the following case studies/projects for better understanding of the chapter.

1. Applications of isomorphism og Graphs for resolving real life problems.
2. How the concept of the connectivity of graphs and cutset play vital role during the COVID-19 pandemic?
3. Draw the map of your country having different states and find the region chromatic number of that map with suitable assumptions. Explain any two applications of the pi coloring of graphs.
4. Coloring of graphs and its applications.
5. Applications of the shortest path algorithms.
6. Applications of graph theory in Pandemic or epidemic.

Exercise

1. How many vertices are there in the graph H if it involves

   (i) 20 edges and degree of each vertex is 2.
   (ii) 35 edges, 5 vertices are of degree 4 and others each of degree 5.
   (iii) 100 edges and degree of each vertex is either 4 or 5.

2. Construct two examples of each type.

   i. Complete graph
   ii. Regular graph

iii. Bipartite and regular graph
iv. Regular but not complete
v. Complete tripartite and regular
vi. Regular, complete, and bipartite graph

3. Explain applications of graph theory in chemistry.
4. Explain any two examples of non-isomorphic graphs having same degree sequences.
5. Take two graphs K_5 and $K_{2,2}$ having some common vertices and edges and find their union, intersection, ring sum, and product of graphs.
6. Show that the Petersen graph is non-planar.
7. Explain any two examples of the following each case graphs.

   i. Graphs having equal maximum degree of a vertex, vertex connectivity, and edge connectivity.
   ii. Vertex connectivity = edge connectivity
   iii. Edge connectivity = Vertex connectivity + 2

8. Give any two examples of the following graphs.

   i. Eulerian and Hamiltonian graphs
   ii. Eulerian but not Hamiltonian graphs
   iii. Hamiltonian but not Eulerian graphs
   iv. Neither Hamiltonian not Eulerian graphs

9. Find the vertex chromatic number, edge chromatic number, region chromatic number, total chromatic number and pi chromatic number of the following graphs.

   i. Wheel graph on 5 Vertices
   ii. Product of C_2 and P_2
   iii. Product of N_2 and P_2

10. Construct an example of weighted graph having 12 vertices and apply Dijkstra's Algorithm to find the shortest path between vertices.
11. Find the adjacency matrix and incidence matrix of the Product of C_3 and P_2.
12. Prove that every planar graph is four colorable.

References

1. Harary F (1969) Graph theory. Addison-Wesley Publishing Company, Reading
2. West Douglas B (2000) Introduction to graph theory, 2nd edn. Pearson Education India, 470 pages. ISBN 13: 9780130144003, ISBN: 9788178088303
3. Bhapkar HR, Mahalle PN, Dhotre PS (2020) Virus graph and COVID-19 pandemic: a graph theory approach. In: Hassanien AE, Dey N, Elghamrawy S (eds) Big data analytics and artificial intelligence against COVID-19: innovation vision and approach. Studies in Big Data. Springer, p 78

4. Bhapkar HR, Mahalle PN, Dey N et al (2020) Revisited COVID-19 mortality and recovery rates: are we missing recovery time period? J Med Syst 44:202. https://doi.org/10.1007/s10916-020-01668-6
5. Bhapkar HR, Mahalle PN, Shinde GR, Mahmud M (2021) Rough sets in COVID-19 to predict symptomatic cases. In: Santosh K, Joshi A (eds) COVID-19: prediction, decision-making, and its impacts. Lecture Notes on Data Engineering and Communications Technologies, vol 60. Springer, Singapore. https://doi.org/10.1007/978-981-15-9682-7_7
6. Bhapkar HR (2018) Applications of planar graph to key cryptography. Int J Pure Appl Math 120(8):89–97. https://acadpubl.eu/hub/2018-120-8/1/10.pdf
7. Thakare S, Bhange A, Bhapkar HR (2024) Graph coloring, types and applications: a survey. Int J Math Combinatorics 1(1):110139
8. Bhapkar HR, Salunke JN (2015) Proof of four colour map theorem by using PRN of graph. J Bull Soc Math Services Standards 3(2):35–42
9. Bhange AA, Bhapkar HR (2020) Perfect coloring of corona product of cycle graph with path and null. Adv Math Sci J 9:10839–10844. https://doi.org/10.37418/amsj.9.12.68
10. Bhange AA, Bhapkar HR (2019) α, β colouring of graphs and related aspects. J Emerging Technol Innov Res 6–4:661–663
11. Bhapkar HR, Bhange AA (2022) PCCP for circular graph with a fan graph. South East Asian J Math Math Sci 18(1):363–372
12. Thakare S, Bhapkar HR (2023) Incident vertex π-coloring of graphs. Commun Math Appl 14(2):591–604. https://doi.org/10.26713/CMA.v14i2.2215
13. Aigner M (2023) Discrete mathematics. American Mathematical Society
14. Erciyes K (2021) Discrete mathematics and graph theory. Springer, Cham. https://doi.org/10.1007/978-3-030-61115-6
15. Rosen K, Krithivasan K (2021) Discrete mathematics and its applications. Mc Graw Hill India. ISBN-978930727353
16. Jiri M, Jaroslav N (2011) Invitation to discrete mathematics, 2nd edn. Oxford University Press. ISBN-9780198570424
17. Tremblay JP, Manohar R (2023) Discrete mathematical structures with applications to computer science. McGraw Hill Education. ISBN-9780074631133

Chapter 10
Trees

> *A common man cannot see the same trees that a mathematician can see*

Abstract This chapter deals with one of the important types of graphs which is derived from nature. We are aware that many concepts of science, technology, social science, philosophy, etc. originate from the various activities of the nature. In nature, one of the important entities is the tree. A common man cannot see the same trees that a mathematician can see. This chapter demonstrates the concepts of trees, properties of trees, traversal trees, spanning trees and minimum spanning trees. Every concept of this chapter is defined by using the BEIMI philosophy. This chapter provides real-life examples for the better understanding of the concepts. It is focused on finding the shortest path among vertices by using various algorithms. The chapter concludes with case studies and exercises aimed at connecting theoretical relational structures to real-world applications across various disciplines.

Keywords Tree · Centre · Spanning tree · Shortest path algorithm

10.1 Introduction and Applications of Trees

This chapter deals with one of the important type of graph which is derived from nature. We are aware that many concepts of science, technology, social science, philosophy, etc., are generated from the various activities of the nature. In nature, one of the important entities is the tree. A common man can see the trees in natural ways, but mathematicians can observe trees in mathematical ways [1].

In 1847, Kirchhoff introduced the concept of trees while studying the various types of electrical networks. Sir Arthur Cayley used the trees to understand and study isomers of saturated hydrocarbons [3]. The trees have numerous applications in various fields such as computer science (sorting hierarchical data, database indexing,

© The Author(s), under exclusive license to Springer Nature Singapore Pte Ltd. 2026

H. R. Bhapkar and P. N. Mahalle, *Discrete Mathematics: An Experiential Learning Perspective*, Studies in Autonomic, Data-driven and Industrial Computing,
https://doi.org/10.1007/978-981-95-7210-6_10

parsing, machine learning, artificial intelligence), cryptography, electrical and electronics engineering, social sciences, mathematical sciences, physical, and chemical sciences [1, 4, 5].

In this chapter, we will study the concept of trees, various types, and applications.

10.1.1 Trees: Experiential Learning

Let's consider the following experiment which is given by the nature.

(A) **Brainstorming Experiment-I**

Let us consider the following pictures of the tree and its skeleton.

(B) **Inferences-I**

In the skeleton of the tree, put a vertex at the point where two branches rise. If two points are adjacent, iff there is a wood between them. Now, observe and understand this tree through mathematical insights. The following are the important inferences of the tree.

1. All points or nodes are connected to each other.
2. There is no circuit or cycle.

(C) **Mathematical Insights-I**

The following are the graph theoretic insights of this experiment.

Trees: A connected and acyclic (Circuit free) graph, is called as a tree. The set of acyclic graphs is called as a forest [1].

It is observed that there are unique trees on one vertex, two vertices, and three vertices.

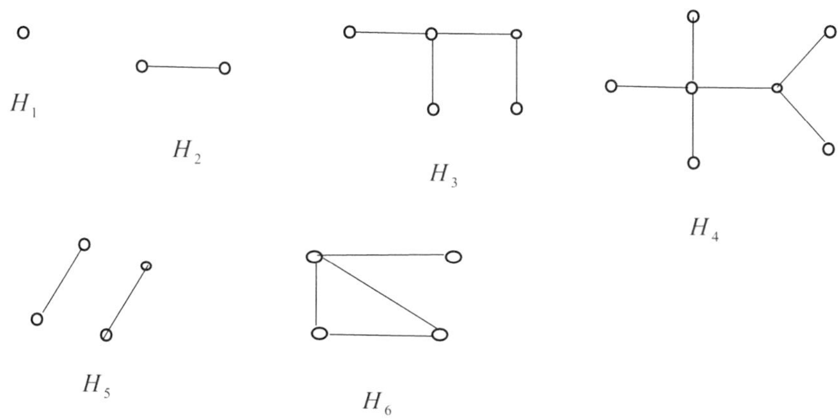

Fig. 10.1 Examples of trees and forest

In Fig. 10.1, the graphs H_1, H_2, H_3, and H_4 are trees. But, the graphs H_5 and H_6 are not trees as H_5 is not connected and H_6 is not circuit free graph. Moreover, the graph H_5 is a set of two trees. Therefore, H_5 is the forest.

A leaf of tree is a vertex having $d(h) = 1$. Other vertices (non-leaf) are known as a branch vertices or internal vertices or branch nodes.

Let's consider the following basic properties of a tree.

Theorem 10.2 A graph H on p vertices is a tree iff H is connected and $|E(H)| = p - 1$.

Theorem 10.4 A tree H with $|V(H)| \geq 2$ has at least 2 vertices of degree 1.

Example 10.1 Draw all non-isomorphic trees (NIT) on 1–6 vertices.

Solution We know that there are unique trees on 1, 2, and 3 vertices. Let's consider the following trees (Figs. 10.2, 10.3 and 10.4).

Example 10.2 Under what conditions trees are complete bipartite graphs.

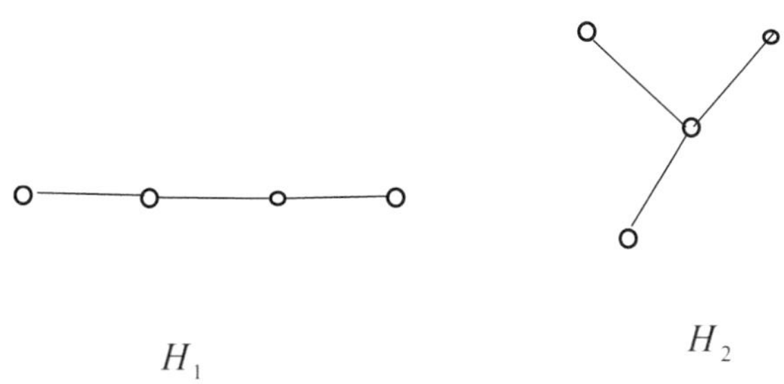

Fig. 10.2 NIT on 4 vertices

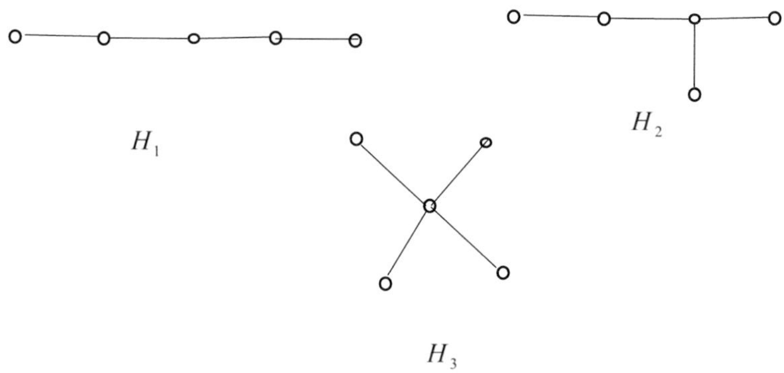

Fig. 10.3 NIT on 5 vertices

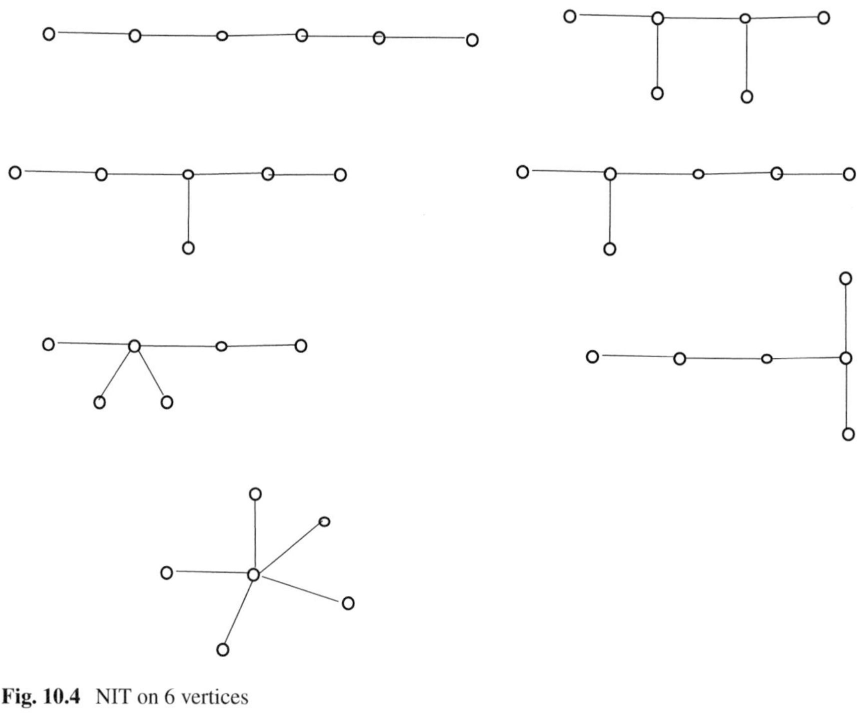

Fig. 10.4 NIT on 6 vertices

Solution Let $H = K_{m,n}$ be a tree.

$$\Rightarrow \left|E(K_{m,n})\right| = m + n - 1 \text{ and } \left|V(K_{m,n})\right| = m + n.$$

Now, as $H = K_{m,n}$, we have

Fig. 10.5 Center of a tree

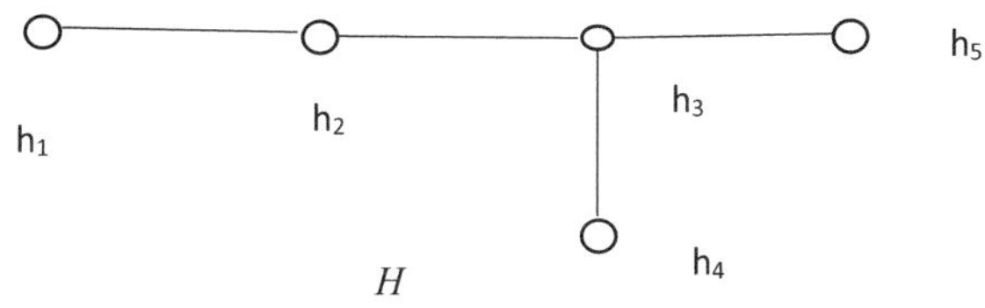

$$mn = m + n - 1 \quad \Rightarrow m = 1 \; or \; n = 1$$

That is, $m = 1$ and any n or $n = 1$ and any m.

Hence, $H = K_{1,n}$ or $H = K_{m,1}$ are only two complete bipartite graphs. These graphs are known as the Star graphs. Thus, H is a star graph.

10.1.2 Center of a Tree

We know that, for any $b, m \in V(H)$,

$d(b, m)$ = length of the shortest path between b and m.

Eccentricity of a Vertex: Let H be a connected graph (CG) and $h \in V(H)$. The eccentricity of a vertex h is denoted by $E(h)$ or $e(h)$ and defined as

$$E(h) = \text{Max}\{d(h, h_i)/\forall h_i \in V(H)\}$$

Center of a Tree

A vertex in a tree having the minimum eccentricity, is called a center of the tree and its eccentricity is known as the radius of a tree. It is denoted as $r(H)$ [3].

Let us consider the following tree H depicted in Fig. 10.5.

Here, $E(h_1) = 3$, $E(h_2) = 2$, $E(h_3) = 2$, $E(h_4) = 3$, and $E(h_5) = 3$,

The minimum eccentricity is 2. So, the tree H has two centers h_2 and h_3.

Theorem 10.1 Every tree H has 1 or 2 centers.

Proof Let H be a tree and $E(h)$ be the eccentricity of the vertex h in H. But, as H is a tree, its eccentricity is attained at the pendent vertices only.

Every tree has at least two pendent vertices. Now, delete all pendent vertices of a tree H and denote a subtree by H_1. Moreover, one edge from each pendent vertex is deleted, so the eccentricity of each vertex in H_1 is reduced by 1. Therefore, the center of H and center of H_1 are same. Continue this process till we get K_2 or K_1. The tree K_2 has two centers and tree K_1 has one center. Hence the proof.

Cut Vertex of a Tree

A vertex whose deletion from the tree H gives a forest is called a cut vertex of the tree H. Every tree has pendent vertices and internal vertices. All internal vertices (non-pendent) are the cut vertices of the tree. In Fig. 10.5, cut vertices are h_2 and h_3.

A connected, circuit free and directed graph is known as directed Tree. A rooted tree T has a vertex $d^-(r) = 0$ and $d^-(i) = 1, \forall\, i \in V(T) - \{r\}$, here d^-: Incoming degree and d^+: Outgoing degree. If $d^-(r) = 0$, then r is the root of T and for a leaf, $d^+(v) = 0$, otherwise, it is internal vertex.

A vertex h of a rooted tree is said to be at level n if $d\ (h_r, h) = n$, where h_r is the root vertex of a tree. The maximum number among levels of all vertices of a tree is called a height of a tree. If two vertices x and y are adjacent and level of x is n and the level of y is $n - 1$, then x is called a child or son of y and y is called a father or parent of x. Moreover, if the levels of two vertices are same then these vertices are called as brothers or sisters of each other.

A rooted tree H in which every interval vertex i has $d^+(i) \leq m$ is a M-ary Tree. A rooted tree H is said to be a binary tree if every internal node of H has at most two sons. A binary tree H is called a full binary tree or regular binary tree if each internal node of H has exactly two sons or zero son.

Figure 10.6 depicts the binary tree and non-binary tree.

In Fig. 10.6, the graph H_2 is full binary tree and H_1 is binary tree.

The maximum of $|V(T)|$ in the full binary tree having m levels is

$$2^0 + 2^1 + 2^{21} + 2^3 + \cdots + 2^m$$

Fig. 10.6 Binary and non-binary trees

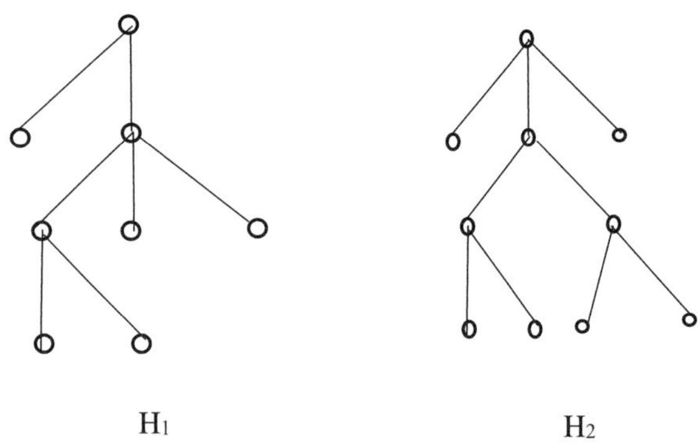

10.2 Tree Traversal or Search: Experiential Learning

The traversing is nothing but visiting or processing all vertices or nodes of a tree according to some rules. The binary tree traversal is the visiting each node of a binary tree only once according to some desirable rules or sequences. There are two types of binary tree traversals according to the nature of the sequence or rules of visiting the nodes [2, 3].

1. **Depth-First Search (Traversal)**

   In this method, the processing starts from the root node (RN) or the most distant descending of the first child. There are two possible to start this process. Therefore, depending upon the sequence, there are three types of Depth-first Search.

   i. **Pre Order Traversal**: In this method, the sequence of visiting the nodes is RN, Left Subtree (LST), and Right Subtree (RST).
      Its sequence is RN-LST-RST (Fig. 10.7).
   ii. **Post-order Traversal**: In this method, the sequence of visiting the nodes is LST, RST, and RN. Its sequence is LST-RN-RST.
      Let's consider the following graph shown in Fig. 10.8.
   iii. **In Order Traversal**: In this traversal, the sequence of visiting the nodes is LST, RN, and RST. Its sequence is LST-RN-RST. Let's consider the following graph shown in Fig. 10.9.

   Now, let's consider the following graph depicted in Fig. 10.10.

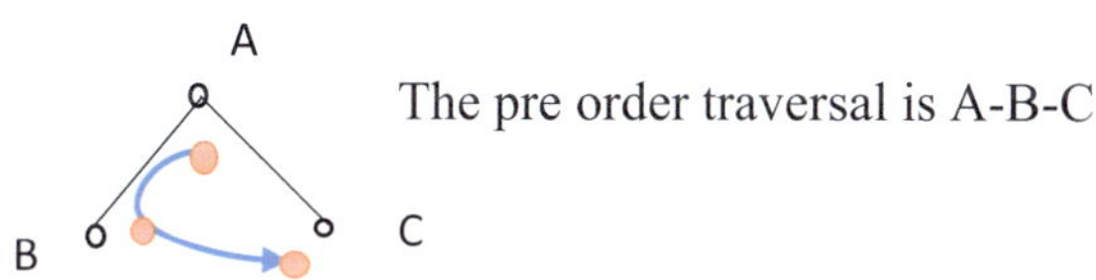

Fig. 10.7 Example of pre-order traversal

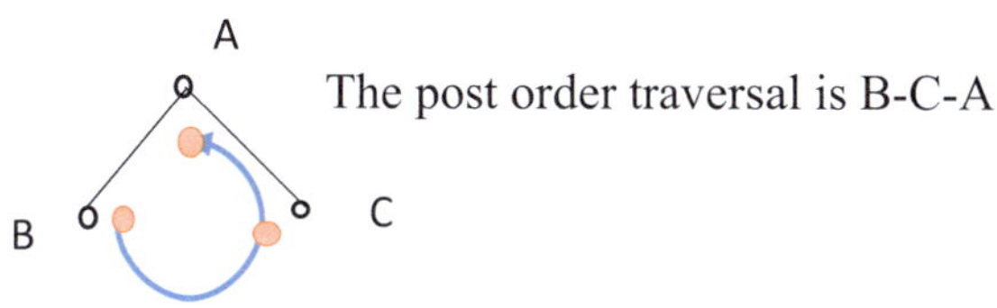

Fig. 10.8 Example of post-order traversal

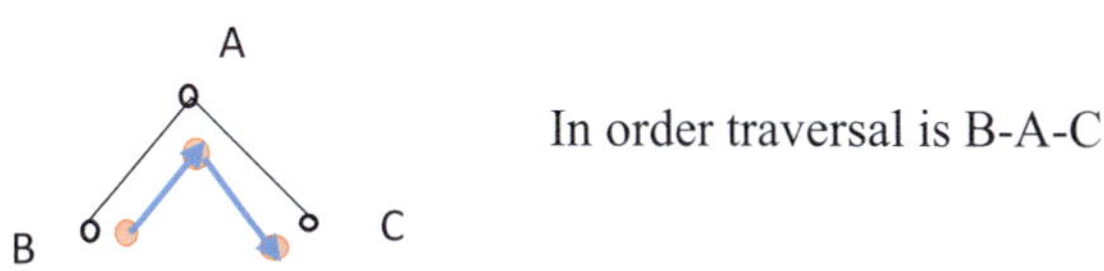

Fig. 10.9 Example of inorder traversal

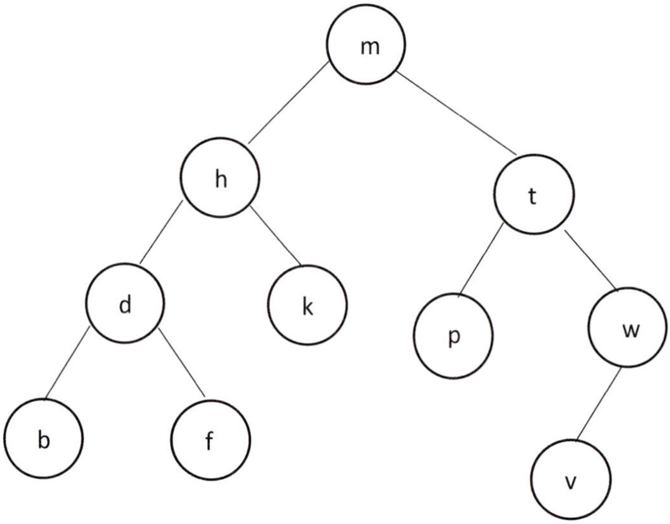

Fig. 10.10 Graph for BFS

(i) Pre-order traversal

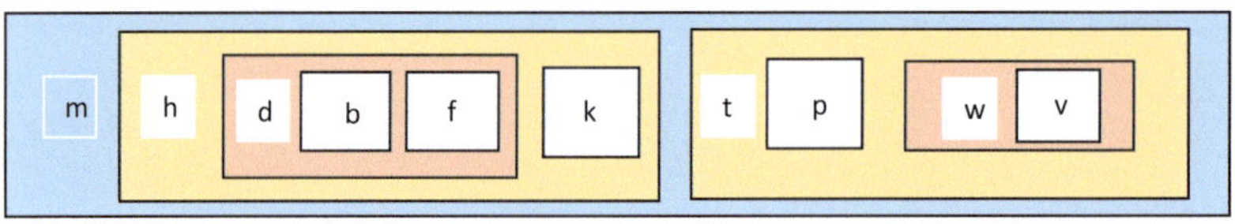

Thus, pre-order traversal is **m h d b f k t p w v**

(ii) Post-order traversal

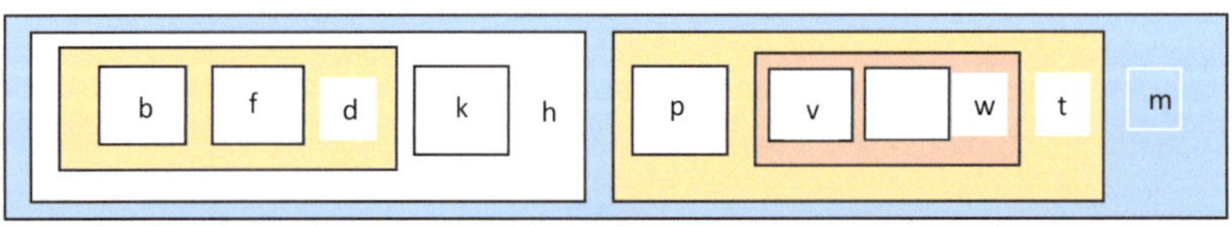

Thus, post-order traversal is **b f d k h p v w t m**

(iii) Inorder traversal

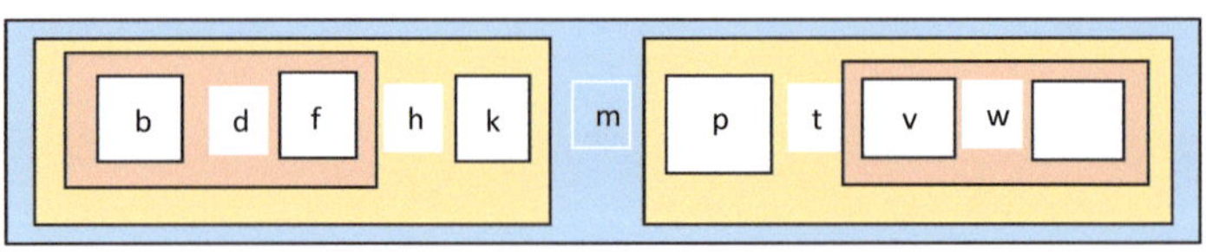

Thus, Inorder traversal is **b d f h k m p t v w**

2. **Breadth First Search or Traversal (BFS)**

In this traversal, the processing starts with the Root node, then level one nodes, level two nodes, level three nodes, and so on, in left to right sequence of nodes, in each level. The root node is at zero level, then children of root node is at level one and so on.

The breadth first traversal of the graph depicted in Fig. 10.10.

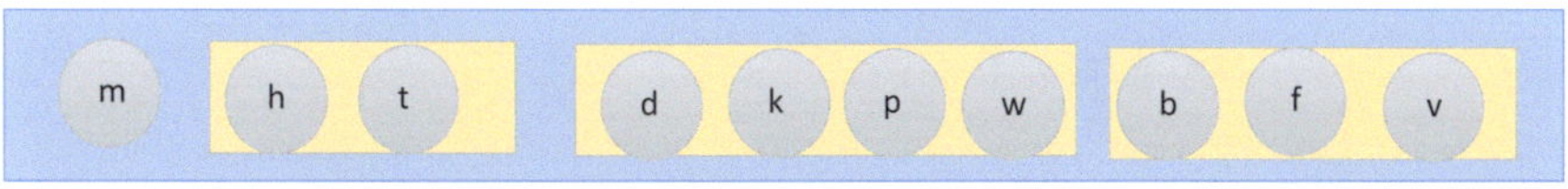

Thus, the breadth first search is **m h t d k p w b f v**

Binary Expression Tree

An algebraic expression can be suitably expressed by using its expression tree. An expression tree is a binary tree with the following properties [2].

(i) Each leaf node of a tree is an operand
(ii) The root and all internal nodes are act as operators
(iii) Subtrees are subexpressions, with the root being an operator.

In this context, we consider only standard arithmetic operations $+$, $-$, *, $\times$, /.

Let's consider the following example depicted in Fig. 10.11.

The expression of given binary tree is

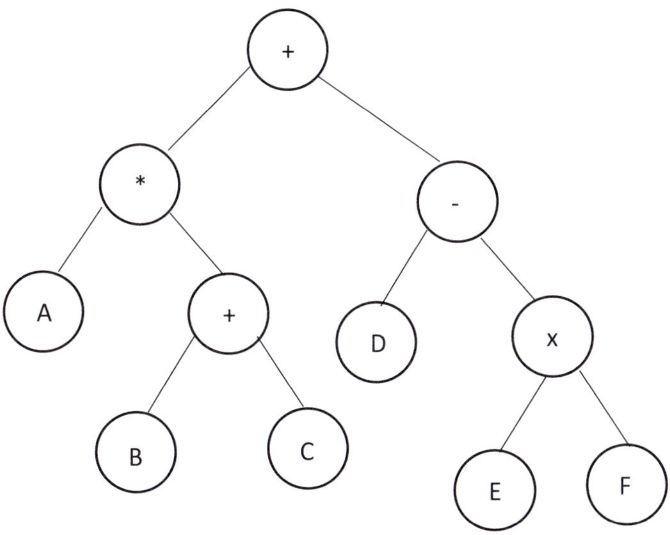

Fig. 10.11 Binary expression tree

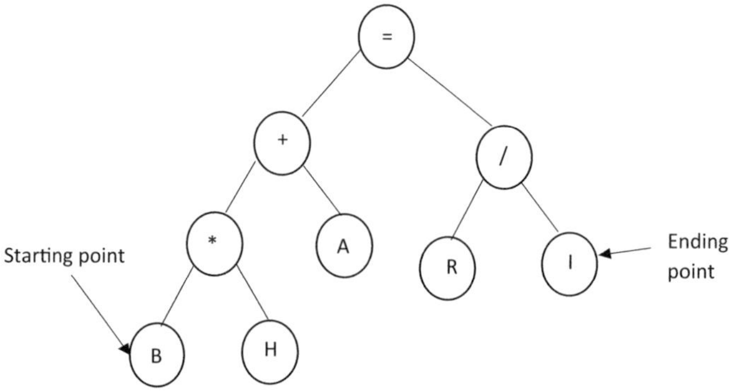

Fig. 10.12 Expression binary tree

$$[A * (B + C)] + [D - (E \times F)]$$

(2) The expression of a binary tree is $[(B * H) + A] = [R/I]$ and the corresponding binary tree is depicted in Fig. 10.12.

Example 10.3 How many internal vertices does a full binary tree with h levels have?

Solution The total number of vertices in full binary tree with h levels

$$= 2^{h+1} - 1$$

Out of which 2^h are leaves.

$$\text{Hence internal vertices} = \left(2^{h+1} - 1\right) - 2^h$$
$$= 2^h(2 - 1) - 1$$
$$= 2^h - 1$$

Example 10.4 What is the total number of nodes in a full binary tree with 20 leaves?

Solution Let n be the number of nodes in a full binary tree.

Then

$$n = mP + 1$$
$$m = 2$$
$$\Rightarrow \quad n = 2P + 1$$

where P is the number of branch nodes.

The number of nodes in a tree is the sum of branch nodes and leaves.

Hence

$$n = P + 20 \Rightarrow P = n - 20$$
$$\Rightarrow \quad n = 2n - 40 + 1 = 2n - 39$$

$$2n - n = 39$$
$$n = 39$$

Hence, the total number of nodes in a full binary tree with 20 leaves are 39

Example 10.5 Can there be a ternary tree with exactly 20 nodes?

Solution We have

$$n = mP + s1$$
$$20 = 3P + 1$$
$$3P = 20 - 1 = 19$$

$P = \frac{19}{3}$ Which is impossible

Therefore, such tree does not exist

Fundamental Circuits

Let H be a connected graph(CG) and a tree $S \subseteq H$ such that $V(S) = V(H)$. If $e \in E(H)$ then e is called as the branch of H. A non-branch edge of H, is a chord of a tree.

The branches and chords of a tree depend upon the selection of the tree of H.

Let S be a spanning tree of a graph H and an edge e in S, is the chord of S. The graph $S + \{e\}$ forms a unique circuit, called as a fundamental circuit of H with respect to tree S. The fundamental circuits depend upon the nature of the spanning trees. Let's consider the following graph and its spanning trees depicted in Fig. 10.13

The graphs S_1 and S_2 are spanning trees of the graph H.

1. For Spanning Tree S_1

The branches of S_1 are e_1, e_2, e_3, e_5, e_7.
The chords of S_1 are e_4, e_6.
Consider the following table of chords and corresponding fundamental circuits of graph.

Chords	Corresponding fundamental circuits
e_4	$\{e_2, e_3, e_4, e_5\}$
e_6	$\{e_1, e_5, e_6, e_7\}$

2. For Spanning Tree S_2

The branches of S_2 are e_2, e_3, e_4, e_6, e_7.
The chords of S_2 are e_1, e_5.
Consider the following table of chords and corresponding fundamental circuits of graph.

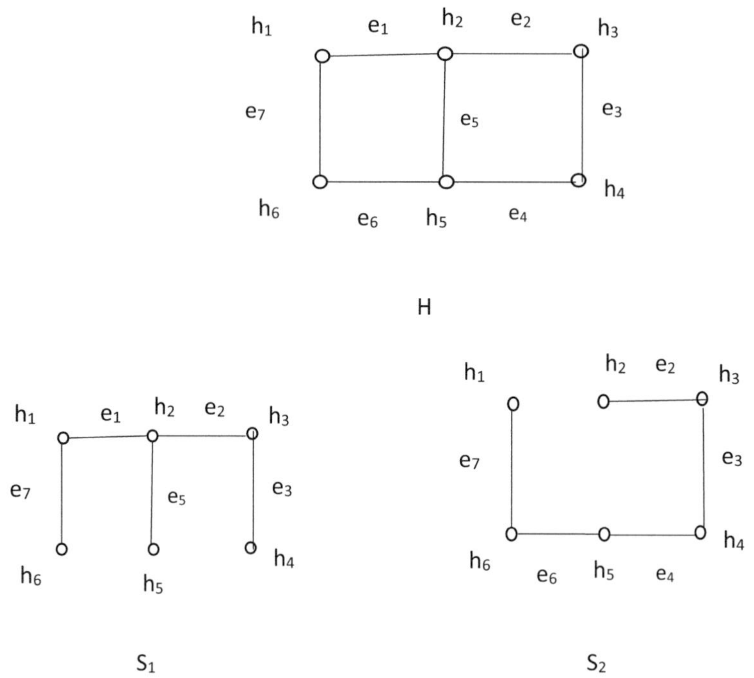

Fig. 10.13 Graph and its spanning trees for fundamental circuits

Chords	Corresponding fundamental circuits
e_1	$\{e_1, e_2, e_3, e_4, e_6, e_7\}$
e_5	$\{e_2, e_3, e_4, e_5\}$

Fundamental Cutsets

Let S be a spanning tree of a connected graph H. Every edge $e \in E(S)$ is an isthmus or bridge of S. The subgraph $S - \{e\}$ divides the tree S into two components say S_1 and S_2, and $V(H) = V(S) = V(S_1) \cup V(S_2)$

An edge set $E_1 \subset E(H)$ joins vertices in $V(S_1)$ to vertices in $V(S_2)$ is a cutset of H. The cutset obtained in this technique is called a fundamental cutset of H with respect to S. The number of fundamental cutsets of a graph is equal to the number of branches of the corresponding tree.

Let's consider the following graphs and its trees depicted in Fig. 10.14. Let H be the graph and S_1 and S_2 be spanning trees of H.

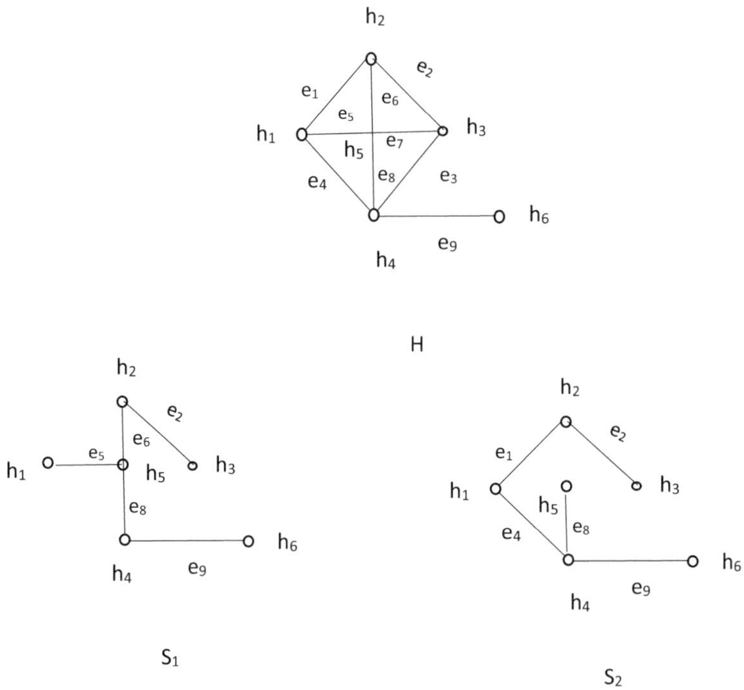

Fig. 10.14 Graph and its trees for fundamental cutsets

(a) For Spanning tree S_1

The branches of S_1 are $e_2,\ e_5,\ e_6,\ e_8,\ e_9$.
The chords of S_1 are $e_1,\ e_3,\ e_4,\ e_7$.
Consider the following table of branches and corresponding fundamental circuits of graph.

Branches	Corresponding fundamental cutsets
e_2	$\{e_2,\ e_3, e_7\}$
e_5	$\{e_1,\ e_4,\ e_5\}$
e_6	$\{e_1,\ e_3,\ e_6,\ e_7\}$
e_8	$\{e_3,\ e_4,\ e_8\}$
e_9	$\{\ e_9\}$

(b) For Spanning tree S_2

The branches of S_2 are $e_1,\ e_2,\ e_4,\ e_8,\ e_9$.
The chords of S_2 are $e_3,\ e_5,\ e_6,\ e_7$.
Consider the following table of branches and corresponding fundamental circuits of graph.

Branches	Corresponding fundamental cutsets
e_1	$\{e_1,\ e_3,\ e_6,\ e_7\}$
e_2	$\{e_2,\ e_3, e_7\}$
e_4	$\{e_3,\ e_4,\ e_5,\ e_6,\ e_7\}$
e_8	$\{e_5,\ e_6,\ e_7,\ e_8\}$
e_9	$\{\ e_9\}$

10.3 Spanning Trees: Experiential Learning

Spanning subgraph is a special type of subgraph of a graph. The spanning subgraph and tree, these two flavors are used together to define the spanning tree of a graph [6, 7].

A subgraph S of a connected graph H is called a spanning tree of H if S is a tree and $V(S) = V(H)$. Let's consider the following examples of spanning trees depicted in the Fig. 10.15.

The subgraphs S_1 and S_2 are connected and circuit free graphs, so S_1 and S_2 are trees. Moreover, S_1 and S_2 have five vertices. Therefore, S_1 and S_2 are spanning trees of the graph H. As subgraph S_3 is cyclic and S_4 is not connected, so S_3 and S_4 are not spanning subgraphs of the graph H.

Notes:

1. For the complete graph K_n, one of its spanning subgraph is a Star graph $K1_{,n-1}$.
2. For the complete bipartite graph $K_{n,m}$, one of its spanning subgraph is a path graph.

10.4 Minimum Spanning Trees: Experiential Learning

Let H be any connected graph. A spanning tree S of a graph H is called a minimum spanning tree (MST) if it's weight is minimum among all existing spanning trees of H. The MST depends upon the minimum weight of the tree, so there may be two or more trees having an equal weight. Thus, the MST may or may not be unique [8].

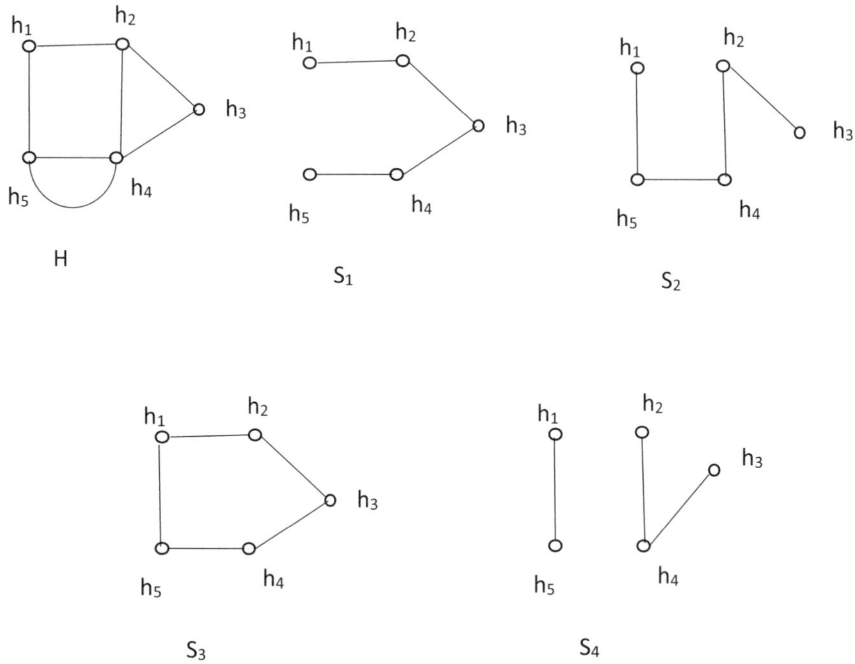

Fig. 10.15 Graph and it's spanning trees

There are various applications of a MST in designing of road networks, railway lines, electric lines, and many more [9–11].

Therefore, it is very important to find the MST of a connected graph. Depending upon, how we select the edges, there are three types of algorithms for constructing the MST.

1. Prim's Algorithm
2. Kruskal Algorithm
3. Bhapkar's Algorithm.

Let's discuss these algorithms one by one.

1. **Prim's Algorithm**

Let be any connected weighted graph. Consider the following steps for designing the MST

Step 1: Select any vertex in H, say $h_0 \in V(H)$. Let S be a tree having vertex set $V(S)$ and edge set $E(S)$. We denote it as $S = \{V(S), E(S)\}$.

Now, initially set $S = \{h_0, \varphi\}$.

Step 2: Find an edge $e_p = \{h_0, h_r\} \in E(H)$ such that it's weight is minimum among all edges whose one end vertex is h_0.

Therefore, set $S = \left\{\{h_0, h_r\}, \left\{e_p\right\}\right\}$.

Step 3: Select an edge $e_q = \{h_s, h_t\} \in E(H)$ such that $h_s \in V(S)$ and $h_t \notin V(S)$ and the weight of e_q is minimum among all edges satisfying this condition. Add the vertex h_t and the edge e_q to the tree S.
Step 4: Repeat step 3 until S contains all vertices of H. The tree S will be the MST

Let us consider the following example of Prim's algorithm.

Example 10.6 Use Prim's algorithm to find MST. Take A = starting vertex (Fig. 10.16).

Solution Consider the following steps for the construction of the minimum spanning tree.

Step 1: Vertex A is starting vertex $S = \{\{A\}, \phi\}$	
Step 2: $S = \{\{A, C\}, e_1\}$	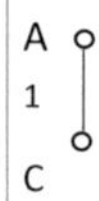
Step 3: $S = \{\{A, C, B\}, \{e_1, e_2\}\}$	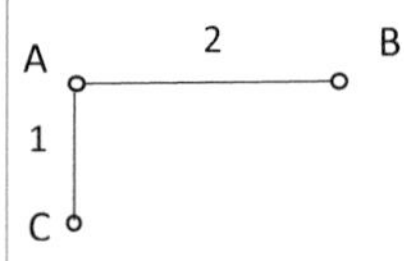
Step 4: $S = \{\{A, C, B, G\}, \{e_1, e_2, e_3\}\}$	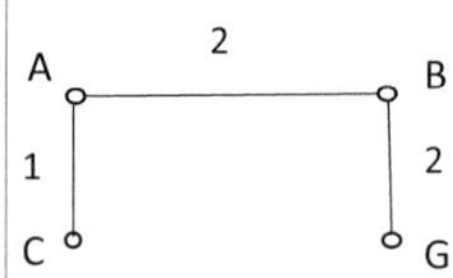
Step 5: $S = \{\{A, C, B, G, H\}, \{e_1, e_2, e_3, e_4\}\}$	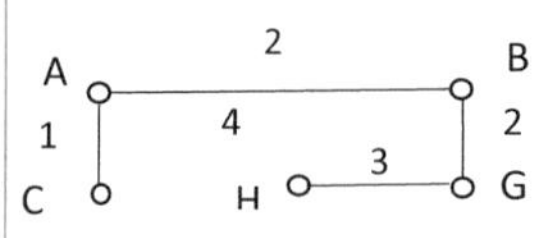

(continued)

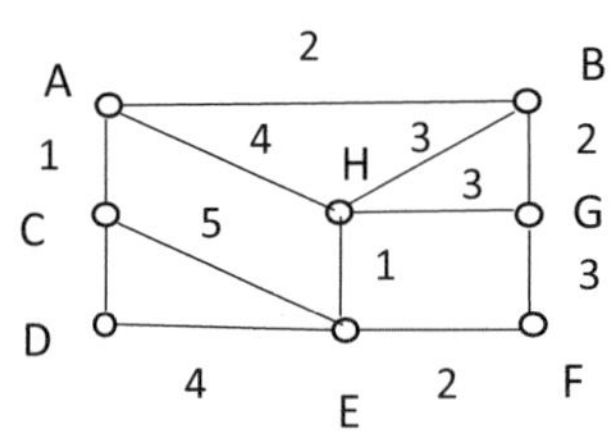

Fig. 10.16 Graph for Prim's algorithm

(continued)

Step 6: $S = \{\{A, C, B, G, H, E\}, \{e_1, e_2, e_3, e_4, e_5\}\}$	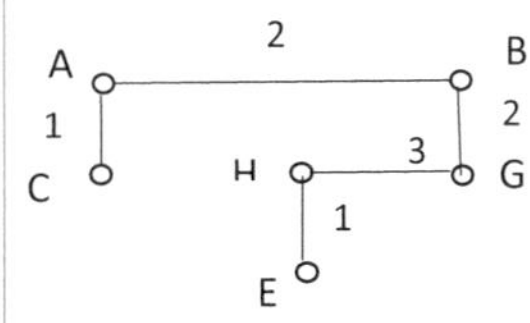
Step 7: $S = \{\{A, C, B, G, H, E, F\}, \{e_1, e_2, e_3, e_4, e_5, e_6\}\}$ The graph obtained is tShe minimum spanning tree of weight $= 1 + 2 + 2 + 3 + 1 + 2 = 11$	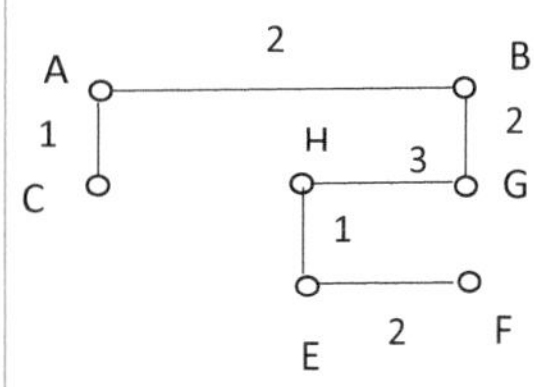

2. **Kruskal's Algorithm**

Let be any connected weighted graph. Consider the following steps for finding the minimum spanning tree [11, 12].

Step 1: Select an edge $e_p = \{h_r, h_s\} \in E(H)$ such that it's weight is minimum among all edges. Therefore, the tree will be $S = \{\{h_r, h_s\}, \{e_p\}\}$.
Step 2: If the edges $e_1, e_2, e_3, \ldots, e_m$ have been chosen then select an edge e_q such that

i. $e_q \neq e_i$ *for* $i = 1, 2, 3, \cdots, m$.
ii. The edges $e_1, e_2, e_3, \cdots, e_m$ and e_q do not form a circuit.
iii. Weight of e_q is minimum among all non-covered edges.

Step 3: Repeat step 2 until S contains all vertices of H. The tree S will be the minimum spanning tree of the graph H [14, 15].

Let us consider the following example of Kruskal's algorithm.

Example 10.7 Use the Kruskal's algorithm to find the MST for the graph shown in Fig. 10.17.

Solution

Step 1: The minimum weight is 2 associated with two edges. These edges do not form a circuit. So select these to edges

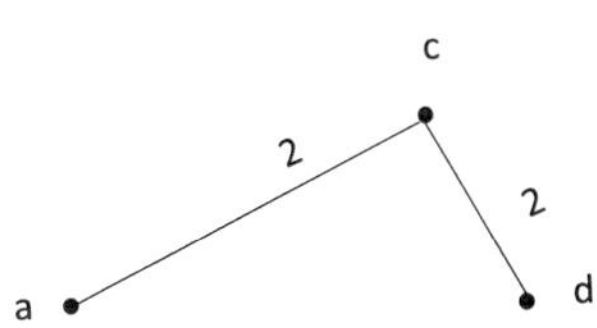

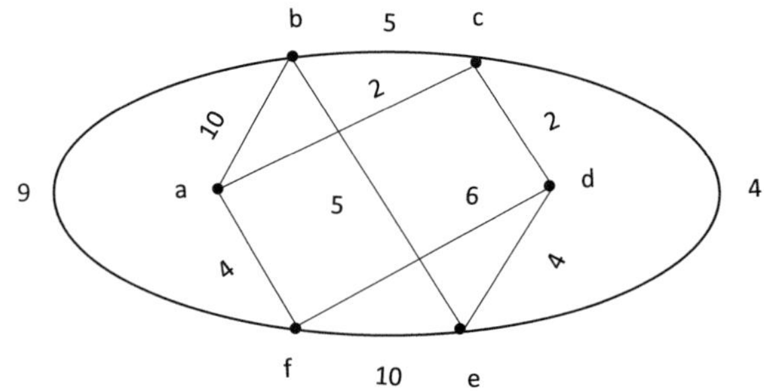

Fig. 10.17 Graph for Kruskal's algorithm

Step 2: Minimum weight = 4

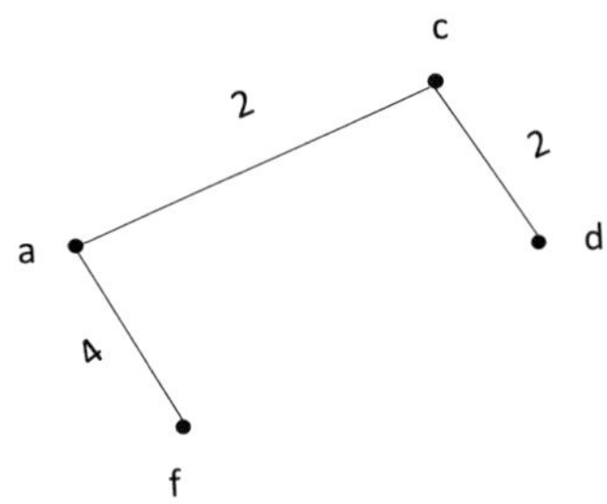

Step 3: maximum weight = 4

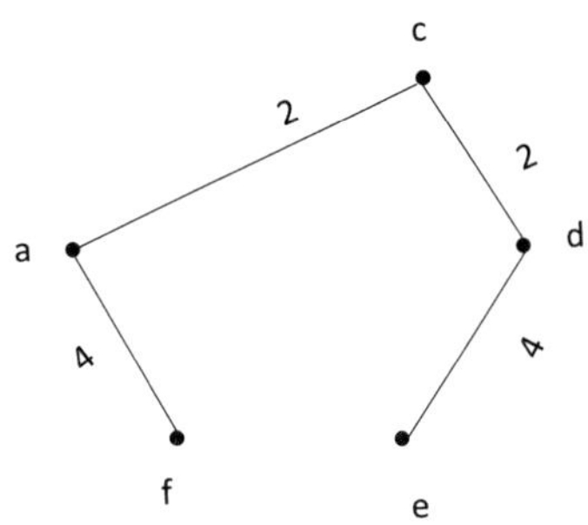

Step 4: Minimum weight = 4 associated with $\{c, e\}$ which forms a circuit so select next minimum weight = 5

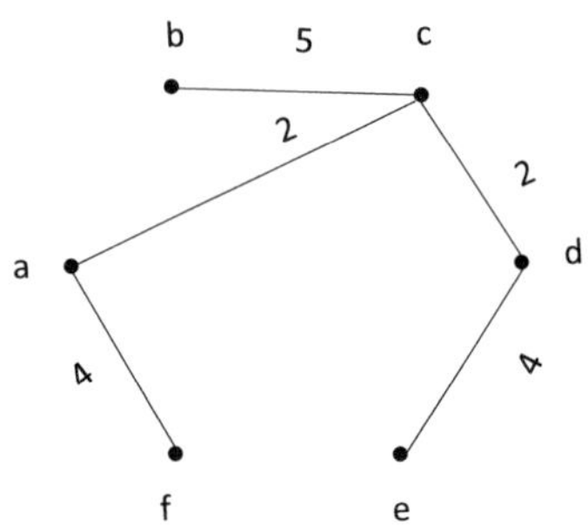

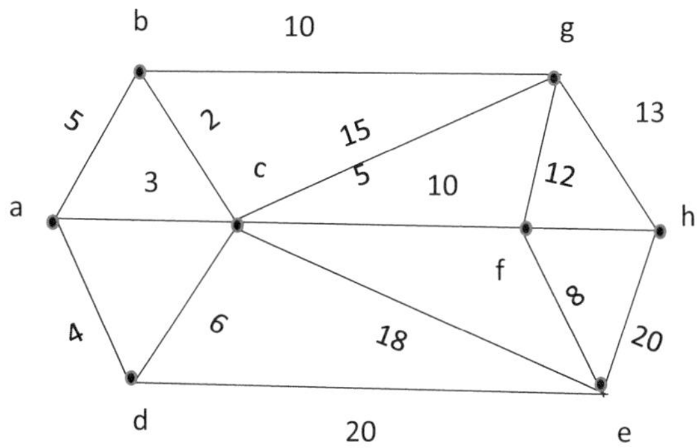

Fig. 10.18 Graph for Bhapkar's algorithm

The graph is the MST and its total weight is 17.

3. **Bhapkar's Algorithm**

Let H (V, E) be a connected weighted graph on n vertices. Consider the following steps to find the MST of a graph H.

Step 1: Form a maximal set H_1, where H_1 is the set of non-adjacent vertices of H. Therefore, $H_1 \subset V(H)$.

Step 2: Let $h \in H_1$, then select an edge incident at the vertex h, having the minimum weight, and do not form a circuit. Find such edges for every vertex in H_1. Let R_1 be the set of end vertices of the selected edges which are not in H_1. Therefore, $S_1 = H_1 \cup R_1$.

Step 3: Repeat steps 1 and 2 for all remaining vertices of graph H, till we get $S_r = V(G)$.

Step 4: If we get forest of subtrees say $T_1, T_2, T_3, \ldots, T_a$ then find the bridges with minimum weights that do not form circuits. Now, connect all such subtrees by the bridges. Thus, we will get the required minimum spanning tree.

The selection of the first set is very important as it is associated with the number of steps required for finding the minimum spanning tree. So, select the set having maximum number of vertices [13].

Let H be a graph having 08 vertices and 15 edges depicted in Fig. 10.18.

Consider the following steps of the Bhapkar's algorithm for finding MST

Step 1: Select H_1, here the set of maximal non-adjacent vertices is H_1.

$$H_1 = \{a, g, h\}$$

Step 2: There are three edges that are incident at the vertex "a." Out of these edges, the $\{a, c\}$ has the minimum weight. Therefore, join the vertex "a" to the vertex c. The weight of an edge $\{g, b\}$ is minimum at the vertex g. So, join g and b by an edge. The weight of an edge $\{e, f\}$ is minimum at the vertex e. Therefore, join e and f vertices. The resultant graph is given below.

$$S_1 = \{a, c, b, g, f, e\}$$

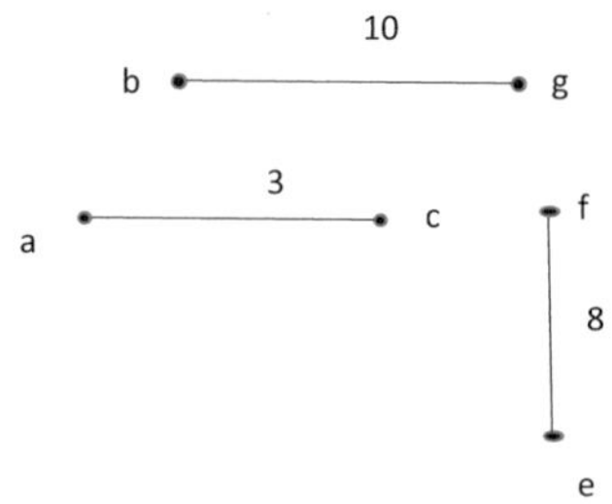

Step 3: In the remaining vertices of the graph H, the set of non-adjacent vertices is H_2 and

$$H_2 = \{d,\ h\}$$

Step 4: The weight of an edge $\{d, a\}$ is the minimum which is adjacent at the vertex d. So, join the vertices d and a. The weight of an edge $\{h, f\}$ is minimum, so join the vertices h and f.

Thus, the resultant graph is given below.

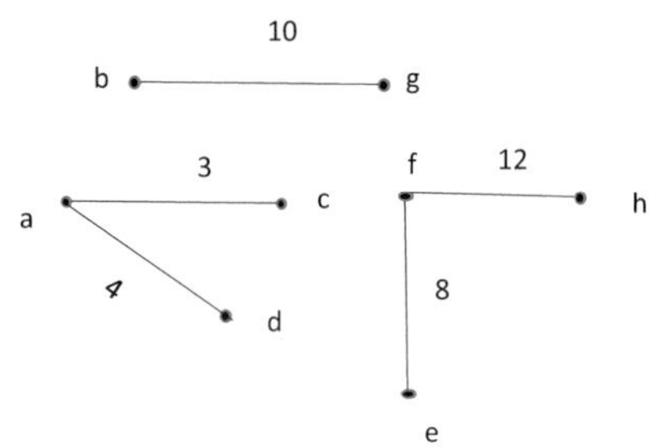

Therefore, $S_1 = \{a, c, b, g, f, e, d, h\}$

Step 5: We have covered all vertices of the graph H. But, we get three components of the tree. Now, connect these three components by the edges having minimum weights. An edge $\{b, c\}$ has the minimum weight. So, join vertices b and c. An edge $\{f, c\}$ has the minimum weight, so join the vertices f and c.

Therefore, the required minimum spanning tree is as follows.

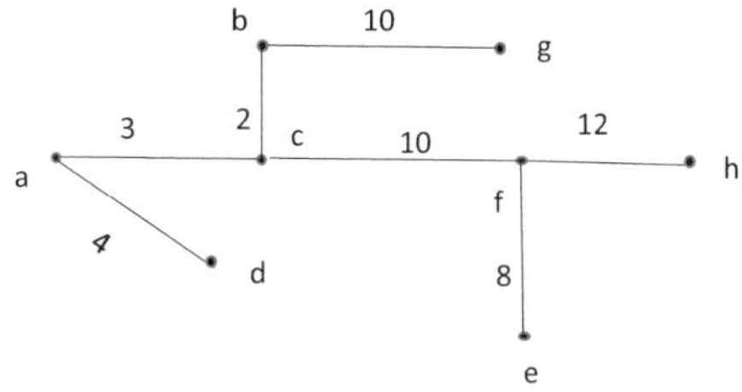

The weight of this spanning tree is $W = 4 + 3 + 2 + 10 + 10 + 8 + 12 = 49$.

10.5 Case Studies and Projects

Let us consider the following case studies/projects for better understanding of the chapter.

1. Applications of trees in real life.
2. Select any two trees around you and draw their graphs. Write degree sequences of the trees. Find center and radius of each tree
3. Find the fundamental cycles and fundamental cutsets of any two graphs having 7 vertices and 11 edges.
4. Binary tree and its applications
5. Minimum spanning tree algorithms and their applications
6. Special report on Bhapkar's algorithm for finding the minimum spanning trees.

Exercise

1. Draw any five trees having at least 5 vertices. Write their degree sequences and comment on their properties.
2. Draw all non-isomorphic trees on

   i. Five vertices
   ii. Six vertices
   iii. Four vertices
   iv. Seven vertices
   v. Three vertices

3. Find the fundamental cycles of the following graphs

   i. Complete graph on 5 vertices
   ii. Wheel graph W_6
   iii. Complete bipartite graph $K_{2,3}$
   iv. Cycle graph C_7

4. Find the fundamental cutsets of the following graphs

   i. K_4
   ii. W_5

iii. $K_{3,3}$
iv. C_7

5. Find pre-order traversal, post order traversal and in order traversal of the trees obtained in examples (3), (i), (iii).
6. Find the breadth first search or traversal of the trees obtained in the example (4).
7. Draw weighted graph on 10 vertices and 15 edges and apply Prim's algorithm to find the MST
8. Apply Kruskal's algorithm to find the MST of a weighted graph 8 vertices, 12 edges, and weight of each edge is different and more than 100.
9. Apply Bhapkar's algorithm to find the MST of weighted graph K_5.
10. Differentiate Prim's, Kruskal's, and Bhapkar's algorithms. Apply these algorithms to find the MST of weighted W_7 graph.

References

1. Harary F (1969) Graph theory. Addison-Wesley Publishing Company, Reading
2. West Douglas B (2000) Introduction to graph theory, 2nd edn. Pearson Education India, p 470. ISBN: 13:9780130144003, ISBN: 9788178088303
3. Rosen K, Krithivasan K (2021) Discrete mathematics and its applications. Mc Graw Hill India. ISBN-978930727353
4. Bhapkar HR, Mahalle PN, Dhotre PS (2020) Virus graph and COVID-19 pandemic: a graph theory approach. In: Hassanien AE, Dey N, Elghamrawy S (eds) Big data analytics and artificial intelligence against COVID-19: innovation vision and approach. Studies in big data. Springer, p 78
5. Bhapkar HR (2018) Applications of planar graph to key cryptography. Int J Pure Appl Math 120(8):89–97. https://acadpubl.eu/hub/2018-120-8/1/10.pdf
6. Graham RL, Hell P (1995) On the history of the minimum spanning tree problem. Ann Hist Comput 7(1):43–57
7. Effanga EO, Edeke UE (2014) Minimal spanning tree of city to city road network. IOSR. J. of Math. 12(4):41–45
8. Arogundade OT, Sobowale B, Akinwale AT (2011) Prim algorithm approach to improving local access network in rural areas. Int J Comp Theory Eng 3(3):413–417
9. Matsui T (1997) A flexible algorithm for generating all the spanning trees in undirected graphs. Algorithmica 18(4):530–543
10. Prim RC (1957) Shortest connection and some generalizations. Bell Syst Tech J 36(6):1389–1401
11. Kruskal JB (1956) On the shortest spanning subtree and the traveling salesman problem. Proc Am Math Soc 7:48–50
12. Yamada T, Kataoka S, Watanable K (2009) Listing all the minimum spanning trees in an undirected graph. Int J Comp Math 4(5):343–348
13. Erciyes K (2021) Discrete mathematics and graph theory. Springer, Cham. https://doi.org/10.1007/978-3-030-61115-6
14. Jiri M, Jaroslav N (2011) Invitation to discrete mathematics, 2nd edn. Oxford University Press. ISBN: 9780198570424
15. Tremblay JP, Manohar R (2023) Discrete mathematical structures with applications to computer science. McGraw Hill Education. ISBN: 9780074631133

Chapter 11
Generating Functions and Recurrence Relations

> *"Without mathematics, there's nothing you can do. Everything around you is mathematics. Everything around you is numbers"*
> —Shakuntala Devi

Abstract This chapter deals with the concepts of generating functions, recurrence relations, related theory and applications. The concepts of this chapter are defined by using examples of computer science. Enumeration problem is a fundamental category of the problem in DM and combinatorics and is solved with the help of generating functions (GF). In other words, GF is a powerful tool to solve various enumeration problems. For a given problem size n, the main objective of the enumeration problem is to determine the number of objects which satisfies certain conditions or definitions of size n. Enumeration problems are used in various applications of DM like algorithm analysis for analyzing the configuration or counting number of inputs in algorithms, predicting or deciding the number of outputs in computation using probabilistic models, possible number of in the encryption process while using encryption and exploring spanning trees using graphs for the purpose of finding paths, cycles, subtrees, etc. The chapter concludes with case studies and projects aimed at connecting theoretical relational structures to real-world applications across various disciplines.

Keywords Generating functions · Reflective observations · Recurrence relation

11.1 Introduction

Enumeration problem is fundamental category of the problem in DM and combinatorics and are solved with the help of generating functions (GF). In other words, GF are the powerful tool to solve various enumeration problems. For a given problem size n, the main objective of enumeration problem is to determine the number of objects which satisfies certain conditions or definitions of size n. Enumeration problems deal with the counting the number of different ways for the purpose of selection,

© The Author(s), under exclusive license to Springer Nature Singapore Pte Ltd. 2026
H. R. Bhapkar and P. N. Mahalle, *Discrete Mathematics: An Experiential Learning Perspective*, Studies in Autonomic, Data-driven and Industrial Computing,
https://doi.org/10.1007/978-981-95-7210-6_11

arrangement or design and is related to the counting theory. [1, 2] Enumeration problems are used in various applications of DM like algorithm analysis for analyzing the configuration or counting number of inputs in algorithms, predicting or deciding the number of outputs in computation using probabilistic models, possible number of in the encryption process while using encryption and exploring spanning trees using graphs for the purpose of finding paths, cycles, subtrees, etc. [3]. There are many examples of enumeration problems in the real world with the operations of listing and counting and few are listed and described below:

- **Example 1**: Consider the example where there are "*n*" number of students in class represented as

$$\mathbf{S} = \{\mathbf{1}, \mathbf{2}, \mathbf{3} \ldots, \mathbf{n}\}$$

From this set, if we would like to know how many permutations are possible for doing the projects in a group then this problem is an enumeration problem and the answer is n!

- **Example 1**: We have data which is represented in the form of 0 and 1 s and represented with the set as

$$\mathbf{D} = \{\mathbf{0}, \mathbf{1}\}$$

If we want to construct the possible codes from the bits of 0 and 1 s, then the number of binary sequences of length will be 2^n [4].

Based on the examples discussed above there are four types of enumeration problems like permutation, combination, counting with repetition and special enumeration problems and are presented in Table 11.1

There are three types of GF like ordinary GF where the sequence of a_n is represented as

$$\mathrm{GF}(x) = \sum_{n=0}^{\infty} a_{n} x^{n}$$

Another category of GF is exponential GF where factorial weights are used for encoding and are represented as below:

$$\mathrm{GF}(x) = \sum_{n=0}^{\infty} \frac{a_{n} x^{n}}{n!}$$

The last category of GF is moment GF where the probability theory and stochastic principles are used to encode the sequence [5].

In general, there are 4 techniques for solving enumeration problems which incudes inclusion–exclusion principle, generating functions, recurrence relations, and combinatorial formulas are depicted in Fig. 11.1.

Table. 11.1 Types of Problems and Examples

Sr. No.	Type of problem	Example
1.	Permutation	– Permutation on the set {*P*, *Q*, *R*}—*PQR*, *PRQ*, *QPR*, *QRP*, *RPQ*, *RQP* – Number of permutations on the set of n objects are n!
2.	Combination	– Combinations of 2 elements from set {*P*, *Q*, *R*} are {*P*, *Q*}, {*Q*, *R*}, and {*P*, *R*} – Total combinations of r objects selected from the set of n objects is given by $(n \text{ Cr}) = n! / (n - r)! * r!$
3.	Counting with repetition	– Stars and bars theorem is used to solve counting with repetition problems – Example—What are different ways of distributing 5 similar biscuits to 3 different kids
4.	Special enumeration problems	– Arrangement of objects with some pre-defined constraints – Dealing with graph data structure for counting of labelled or un-labelled objects –

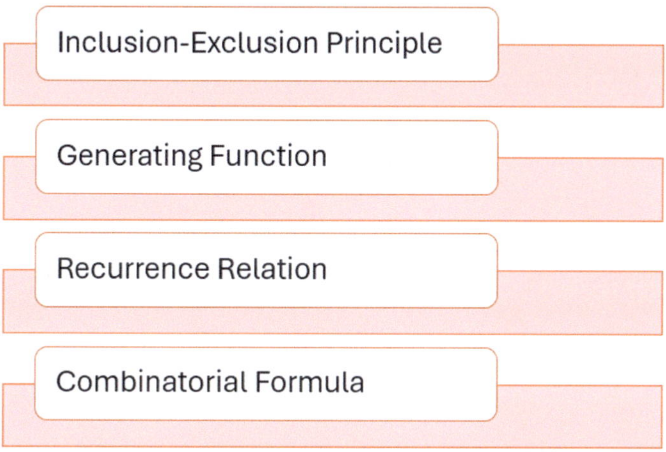

Fig. 11.1 Techniques for solving enumeration problems

1. **Inclusion–Exclusion Principle**

In this method, intersections are added and subtracted for finding number of objects in overlapping sets.

2. **Generating Function**

In this method, a systematic approach of designing a GF is employed for encoding series or sequence for solving enumeration problems.

3. **Recurrence Relations**

In this method, a computation formula is design in the form of recurrence relation to generate set of series which is finite or infinite in nature. This recurrence relation is then solved iteratively or GF.

4. **Combinatorial Formulas**

In this method, standard formulas of permutations and combinations are used to find solutions to the simple problems.

In the scope of this chapter, generating functions and recurrence relations are described in detail using experiential learning.

11.2 Generating Functions—Experiential Learning

GF is one of the most popular techniques to solve enumeration problems. Modeling of sequence of numbers can be done more effectively using GF. For a given enumeration problem of size *n*, GF mainly deals with mainly generating sequence of numbers denoted as a_n [3].

Let (a_n) where $n \geq 0$ is a sequence of numbers denoted. The GF to generate this sequence is given by

$$A(x) = \sum_{n\geq 0}^{n} a^n x^n$$

If we are interested in generating the class of binary sequence where the length of the sequence is equal to size of sequence. In this case, GF is given as

$$A(x) = \sum_{n\geq 0}^{n} 2^n x^n$$

As there are $a_n = 2^n$ binary sequences for the problem size of n.

To understand GF using learning by doing philosophy of EL, the tools like direct experience, real-world applications and reflection are used. EL involves several hands-on activities and solving the real-world problems with more efficient solutions. As mentioned in [1], Kolb's EL theory states that there are main four stages of EL and are depicted in Fig. 11.2.

1. **Concrete Experience**

In this stage, the learner is involved in getting a specialized experience using direct participation in the activity enabling learning by doing. The more focus is on real-world business cases and practical exposure.

Fig. 11.2 Stages of EL

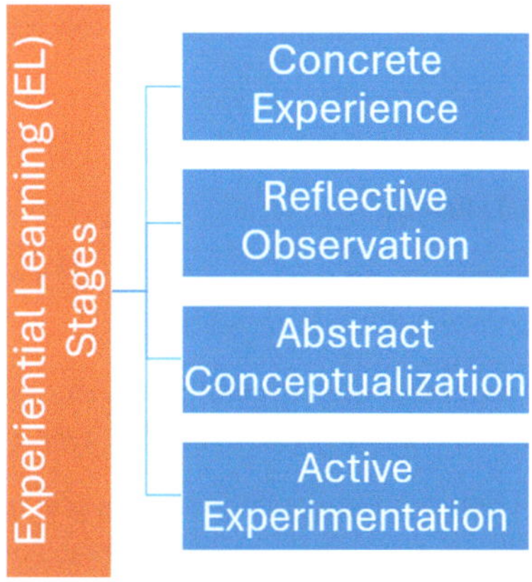

Example: A team of faculty or doing some group activity in the leadership program where there is involvement of every faculty to gain specialized experience on the development of leadership.

2. **Reflexive Observation**

Once the learner gets some experience using learning by doing, the step back is performed to see what has happened. Various perspectives are used to get more insights on the observations and experiences.

Example: A team of faculty put more insights into the key challenges faced during the activity, group cohesion and the main outcomes.

3. **Abstract Conceptualization**

Formulation of topical theory or concept on the reflections received is earlier stage is carried out by learners in this stage. The mapping of learners' experience with either framework or principles is established in this stage by the learners.

Example: A faculty team envisages the team building principles or respective strategic decisions that will help to improve future decision-making processes.

4. **Active Experimentation**

In this stage, learner apply recently gained knowledge or insights to new use cases or business cases. This stage needs planning to apply acquired knowledge to explore new applications.

Example: Faculty team apply the acquired knowledge at new leadership positions to witness better outcomes.

Consider an interesting example of explaining and teaching GF to the group of students.

Objective: Objective of this activity is to use GF to solve category of combinatorial problem. Consider the scenario where we have some amount "x" and we are interested to find the various possible ways to create a change for some amount with the help of coins.

Actors: Students

EL Steps:

Concrete experience: This stage will involve counting the coins for possible combinations to make a change of "*y*" amount by students.

Reflective observation: Students will encounter various challenges in counting the coins manually and will conclude using GF to expedite and simplify the process.

Abstract conceptualization: The main aim of this stage is to design GF for the above-mentioned problem of the form as given below:

$$\left(1 + x + x^2 + \cdots\right)\left(1 + x^5 + x^{10} + \cdots\right)$$

Active experimentation: In this stage, the above GF is used to solve different types of combinatorial problems with varies limits and constraints.

11.3 Recurrence Relations—Experiential Learning

Recurrence relation (RR) is a computational rule or equation used to generate series or set which is finite and/or infinite in nature. RR defines the sequence of values expressed in terms of collection of the preceding terms. As the name indicates, RR is represented with the recursive function where each term is defined in terms of the previous term. RR plays a crucial role in design and analysis of algorithms. RR can be designed for every algorithmic strategy and every recursive algorithm with the problem size of "n" can be modeled with RR to decide the cost or runtime complexity of the algorithm. Combinatorics is another area of computer science where RR is used to find out solutions to counting problems. Examples include different possible ways to arrange students in the queue or construction of binary trees given number of "n" nodes, etc. The classical example of RR is to generate the Fibonacci series where every term is calculated recursively using the previous two terms [3, 5].

The general form of RR is

$$a\_n = f(a\_0, a\_1, \ldots, a\_\{n - 1\})$$

There are main three techniques used to solve RR and are described in Fig. 11.3.

- **Iteration**: There is repeated application of RR to find or generate next term of series or sequence in iterative way of finding solution to RR. This repetition is stopped when all the required terms are generated.

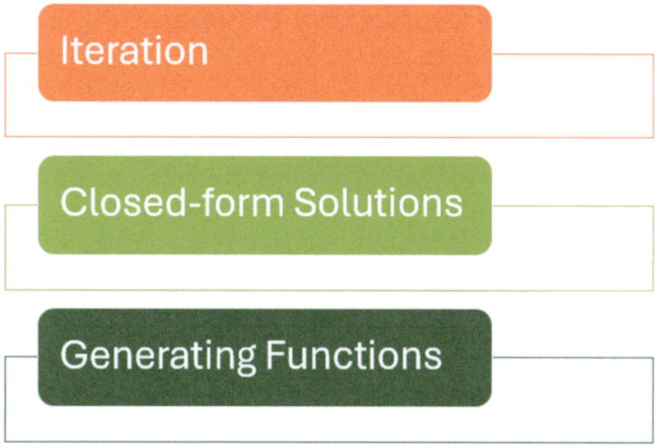

Fig. 11.3 Techniques to solve recurrence relations

- **Closed-form solutions**: In this method, the formula is modeled to find the last term in the sequence without using the preceding terms.
- **Generating functions**: GF is designed to solve RR.

Integration of EL to deliver RR will help to improve the learning curve and in turn will also result in productive student engagement. This integration also helps to improve technology readiness level of the learner. In view of this, according to Kolb's stages of learning in EL, teaching RR will work as follows:

- **Concrete experience**

In this stage, the task is given to students to think of real-world problems or example which can be solve with an application of RR. Consider the example of designing RR for divide and conquer algorithmic strategy.

- **Reflective observation**

In this stage, there is a exploration on selection of one of the RR solving techniques to solve RR. This can be done by observing patterns or association between the terms of the sequence.

- **Abstract conceptualization**

The school of thought on application of RR to analyze algorithms in terms of cost and running time complexity is carried out in this stage. Discussion on how similar philosophy can be used to design RR of various algorithms in engineering and applied sciences.

- **Active experimentation**

The actual use of RR on complex problems in engineering and sciences is carried out in this stage. Analysis of different RR for respective problems toward varied patterns and behaviors is also the part of this stage.

11.4 Case Studies and Projects

The following are the case studies/projects of this chapter.

1. Applications of generating functions in cryptography
2. Recurrence relation in social sciences
3. Generating functions and recurrence relations in Engineering

References

1. Morris TH (2020) Experiential learning–a systematic review and revision of Kolb's model. Interact Learn Environ 28(8):1064–1077
2. Bona M (2012) Combinatorics of permutations, 2nd edn. CRC Press. ISBN: 978-1-4398-5051-0
3. Rosen K, Krithivasan K (2021) Discrete mathematics and its applications. Mc Graw Hill, India. ISBN: 978930727353
4. Jiri M, Jaroslav N (2011) Invitation to discrete mathematics, 2nd edn. Oxford University Press. ISBN: 9780198570424
5. Tremblay JP, Manohar R (2023) Discrete mathematical structures with applications to computer science. McGraw Hill Education. ISBN: 9780074631133

Chapter 12
Algebraic Structures with One Binary Operation

No human investigation can be called real science if it cannot be demonstrated mathematically.
—Leonardo Da Vinci

Abstract The group theory was developed in the nineteenth century to solve equations. The groups play an important role in developing various applications in Mathematics, computer sciences, cryptography, medical sciences, biology, and many other related fields. For a better understanding of the groups, an imagination of the learners must be righteous. This chapter deals with the various operations, groups, subgroups, cyclic group, normal subgroup, homomorphism, isomorphism, and group codes with precise examples. The chapter defines the concepts of group and related theories by using the experiment of the common family BHAAHA having three members. It explains how group theory connects with regular family activities. Moreover, due to the various applications in Artificial Intelligence and Machine Learning, group theory is in the limelight of scientists of various fields. The chapter culminates in case studies, projects and pedagogical exercises aimed at reconciling theoretical relational frameworks with their multidisciplinary applications.

Keywords Group · Cyclic group · Isomorphism of groups · Group codes

12.1 Introduction

The group theory was developed in the nineteenth century to solve equations. The groups play an important role in developing various applications in Mathematics, computer sciences, cryptography, medical sciences, biology, and many other related fields. For a better understanding of the groups, an imagination of the learners must be righteous. Due to the various applications in artificial intelligence and machine learning, group theory is in the limelight of scientists of various fields [1].

This chapter deals with the various operations, groups, subgroups, cyclic group, normal subgroup, homomorphism, isomorphism, and group codes [1, 2].

© The Author(s), under exclusive license to Springer Nature Singapore Pte Ltd. 2026

H. R. Bhapkar and P. N. Mahalle, *Discrete Mathematics: An Experiential Learning Perspective*, Studies in Autonomic, Data-driven and Industrial Computing,
https://doi.org/10.1007/978-981-95-7210-6_12

12.2 Binary Operations and Properties: Experiential Learning

(A) **Brainstorming Experiment-I**

Let us consider the following real story. There is a family named BHAAHA living together. They lived happily together, having three members in a family: father, mother, and their son. The family has narrative, courage, resilience, love, compassion, joy, difficulty, and many more emotions and reactions. Master Atharva is living with his parents, Hari and Archana. All family members love each other and enjoy their lives. Hari and Atharva are working at the university and a software company, respectively, but Archana is supporting Hari and Atharva. When she is alone, she enjoys her hobbies, but if she is with Hari and Atharva, then she helps them in their work. If Hari is alone, then he works or thinks about the career of Atharva, but if he is with his son, then they together help in the work of mother. If the son is alone, then he does work that will help his father. All family members are blessings for each other and also support systems for one another.

Let, $H = \{M, F, S\}$ and the operation * defined as $h * m = p$ iff h and m are together then both help in the work of p or both think about p.

The tabular representation of the above story is as follows.

*	M	F	S
M	M	F	S
F	F	S	M
S	S	M	F

(B) **Inferences-I**

The following are the brilliant inferences of Experiment-I

1. All elements of the above table are in the set H (Binary Operation).
2. As this table is symmetric, so, commutative.
3. The operation is associative.
4. The mother is a very helpful member in the family. So, M is the identity element.
5. Father and son are inverses of each other.

(C) **Mathematical Insights-I**

The following are some important mathematical insights of this experiments.

(I) **Binary Operation (BO)**

Let H be any non-empty set. A function $f : H \times H \rightarrow H$ is called the binary operation. Moreover, '$*$' is a binary operation or binary function on the set A iff $h_1 * h_2 \in H, \;\; \forall \;\; h_1, h_2 \in H$ and $h_1 * h_2$ is unique [1].

An n – ary operation: It is $f : H \times H \times \cdots \times H \rightarrow H$

n=1	"Unary Operation"
n=2	"Binary Operation"
n=3	"Ternary Operation"

Let us consider the following table having sets and the binary operations.

Sets	Usual operations			
	Addition	Subtraction	Multiplication	Division
Complex numbers	Yes	Yes	Yes	No
Real numbers	Yes	Yes	Yes	No
Non-zero real numbers	Yes	Yes	Yes	Yes
Rational numbers	Yes	Yes	Yes	No
Integers	Yes	Yes	Yes	No
Natural numbers	Yes	No	Yes	No

In the table, "Yes" stands for the binary operation with respect to the corresponding set and the operation and "No" stands for not binary operation.

(II) **Properties of binary operations**

Let $H \neq \phi$ be any set.

1 **Commutative property**

A binary operation '*' on H is said to be commutative if

$$h_1 * h_2 = h_2 * h_1 \; ; \; \forall \quad h_1, h_2 \in H.$$

'+' and 'x' are commutative binary operations on $\Re$.
But '_' is not commutative operations on $\Re$.

2 **Associative Property**

A binary operation '*' on H is associative if

$$h_1 * (h_2 * h_3) = (h_1 * h_2) * h_3 \; ; \; \forall \quad h_1, h_2, h_3 \in H.$$

e.g., '+' and 'X' are associative, but '_' is not associative on $\Re$.

3 **Existence of the Identity Element**

Let '*' binary operation on the non-empty set H. An element $e \in H$ is identity element with respect to * if $h * e = e * h \; ; \; \forall \quad h \in H$.

4 **Existence of the Inverse Element**

If $h \in H$ and $\exists \; m \in H$ such that

$h * m = m * h = e$ then m the inverse of h in H with respect to *. It is denoted by $h^{-1} = m$.

Example 12.1 For each of the following, determine whether * is a binary operation. If * is binary then check which are commutative, associative and find e and h^{-1}

(i) $\Re$ & $h * m = h \times m$.
(ii) Z^{+} & $h * m = \frac{h}{m}$.
(iii) On Z^{+}, where $h * m = m - 100 - h$.
(iv) On $\Re$, where $h * m = \text{Min}\ (h, m)$.
(v) On $\Re$, where $h * m = h \times |m|$.
(vi) On Z, where $h * m = h^{m}$.

Solution

(i) The operation is binary. It is also commutative and associative. $e = 1$ and $h^{-1} = \frac{1}{h},\ h \neq 0$
(ii) Not binary, not commutative, not associative, no identity element, no inverse element.
(iii) Not binary, not commutative, not associative, no identity element, no inverse.
(iv) Binary operation, commutative, associative, no identity element, no inverse.
(v) Binary operation, not commutative, not associative, no identity element, no inverse.
(vi) Not binary, not commutative, not associative, no identity element, no inverse.

12.3 Spaces Algebraic Structures: Experiential Learning

Let us consider the following brainstorming experiment, its inferences and the mathematical insights.

(A) **Brainstorming Experiment-II**

Consider Experiment-I, to find the brilliant inferences associated to this topic.

(B) **Inferences-II**

The notable inferences are as follows.

i. All elements of the table belong to H, so * is BO on H (Groupoid).
ii. The * is associative on H (Semi Group).
iii. The identity element is Mother, i.e., M (Monoid).
iv. The inverse of M, F, and S are M, S, F, respectively (Group).
v. The * is commutative as the table is symmetric (Abelian Group).

(C) **Mathematical Insights-II**

Now, express the above inferences in terms of the mathematical theory or notations or symbols. The above three inferences suggest that, there are mainly five types of an algebraic structures.

1. Groupoid.
2. Semi Group.

3. Monoid.
4. Group.
5. Abelian Group.

Let's define one by one by using the above inferences and experiment.
In algebraic structure $(H, *)$, $H \neq \phi$ and $*$ is binary operation (BO) on H.

12.3.1 Groupoid

An algebraic structure $(H, *)$ is called a Groupoid if $h * m \in H; \quad \forall\, h, m \in H$.

In the above experiment, $*$ is binary operation on H. Therefore, $(H, *)$ is a Groupoid.

An algebraic structures $(\Re, +)$, $(\Re, -)$,$(\Re, x)$,$(Z, +)$,$(Z, -)$,(Z, x) are Groupoid. But $(\Re, \div)$, $(Z, \div)$, $(N, \div)$, $(N, -)$ are not Groupoid.

12.3.2 Semi Group

An algebraic structure $(H, *)$, is a Semi group if "*" is

1. Binary operation on H.
2. Associative on H.

In the above experiment, $*$ is binary operation on H as well as associative. Therefore, $(H, *)$ is a Semi group.

An algebraic structures $(\Re, +)$, $(\Re, x)$, $(Z, +)$, $Z, x)$ are Semi groups. But $(\Re, \div)$, $(\Re, -)$, $(Z,\div)$, $(Z, -)$, $(N, \div)$, $(N, -)$ are not Semi groups.

12.3.3 Monoid

An algebraic structure $(H, *)$, is called a Monoid if "*" is

1. Binary operation on H.
2. Associative on H.
3. Existence of Identity: $e \in H$ is the Identity element with respect to $*$ if $h * e = e * h\,; \ \forall \ \ h \in H$ [2, 3].

In the above experiment, * is binary operation, associative and the identity element exists in H. Therefore, $(H, *)$ is a Monoid.

An algebraic structures $(\Re, +)$, $(\Re, x)$, $(Z, +)$, (Z, x) are Monoids. But $(\Re, \div)$, $(\Re, -)$, $(Z, \div)$, $(Z, -)$, $(N, \div)$, $(N, -)$ are not Monoids.

Let us consider the following examples which differentiate among Groupoid, Semi group, and Monoid.

Set	Operation	Binary	Associative	Identity	Remark
N: Natural Number set	$h * m = h - m$	No	No	No	Not Groupoid Not Semi group Not Monoid
$\Re$	$h * m = h - m$	Yes	No	No	Groupoid Not Semi group Not Monoid
$\Re^{+}$: Positive real number set	$h * m = h + m$	Yes	Yes	No	Semi group but Not Monoid
$\Re$	$h * m = h + m$	Yes	Yes	Yes	Monoid

Notes:

1. Every Monoid is Semi group. But the converse is not true. The Semi group may or may not be Monoid.
2. Every Semi group is Groupoid, but converse is not true.

12.4 Groups and Abelian Groups: Experiential Learning

Consider Experiment-II and the corresponding inferences.

12.4.1 Groups

An algebraic structure (H, *), is called a Group if

1. Binary operation: "*" is BO on H.
2. Associativity: "*" is associative on H.
3. Existence of the Identity in H.
4. Existence of the Inverse for each $h \in H$ in H [4].

12.4.2 Abelian Group

A group (H, *) is called an Abelian group if * is commutative in G [1, 5]

12.4.3 Properties of Group

Property I The Identity element (IE) in a group is unique.

Proof e_1 is $IE \Rightarrow e_1 * e_2 = e_2$, and
e_2 is $IE \Rightarrow e_1 * e_2 = e_1$, i
$\Rightarrow e_1 * e_2 = e_1 = e_2$, Hence the proof.

Property II In group, if m and p are inverses of h, then $m = p$.

Proof Let h be any element in group $(H, *)$. Let e be the identity element in H. Suppose m and p be two inverses of h in H. Therefore,
$h * m = e = m * h$, and $h * p = e = p * h$
We have,

$$m = m * e = m * (h * p)$$
$$m = (m * h) * p = e * p = p$$
$$m = p$$

Hence the proof.

Property III In group $(H, *)$, $\left((d)^{-1}\right)^{-1} = d, \ \forall\, d \in H$

Proof We have,

$$(d)^{-1} * d = e$$
$$\Rightarrow \left(d^{-1}\right)^{-1} * \left(d^{-1}\right) * d = \left(d^{-1}\right)^{-1} * e$$
$$\Rightarrow \left[\left(d^{-1}\right)^{-1} * \left(d^{-1}\right)\right] * d = \left(d^{-1}\right)^{-1} * e$$
$$\Rightarrow e * d = \left(d^{-1}\right)^{-1}$$
$$\Rightarrow d = \left(d^{-1}\right)^{-1}$$

Hence the proof.

Property IV In group (H,*), $(h * m)^{-1} = m^{-1} * h^{-1};\ \forall\, h, m \in H$.

Proof Let h^{-1} and h^{-1} be the inverses of h and m, respectively, in H.
Consider, $(h * m) * (m^{-1} * h^{-1}) = h * (m * m^{-1}) * h^{-1}$

$$= h * h^{-1} = e \tag{12.1}$$

Similarly,

$$(m^{-1} * h^{-1}) * (h * m) * = e \tag{12.2}$$

From equations (12.1) and (12.2), $m^{-1} * h^{-1}$ is the inverse of $h * m$.

Property V Prove that the cancelation law holds in group.

Proof Let $h, m, p \in H$ and e be the identity element in H.

$$\begin{aligned} &h * m = h * p \\ &\Rightarrow h^{-1} * (h * m) = h^{-1} * (h * p) \\ &\Rightarrow e * m = e * p \\ &\Rightarrow m = p \end{aligned}$$

Similarly, $m * h = p * h \Rightarrow m = p$. Hence the proof.

Property VI In $(H,^*)$ group, $h * x_1 = m$ and $y_1 * h = m$ have unique solutions in H.

Proof Assume there are two solutions of the statement and prove that the 2 solutions are equal.

12.5 Permutation Groups: Experiential Learning

In 7th chapter, we have studied the permutations on n elements and their properties with suitable examples. Let $S = \{1, 2, 3, \cdots, n\}$ be finite set with n distinct elements. Let H be the set of all permutations on n elements of the set S. Let the operation * is the product of two permutations in the set H. The product of two permutations is obtained by procedure of the composition of two mappings. Therefore, the operation * is binary on the set H.

Identity Element = Identity Function and Inverse Element = Inverse Function. Therefore, if

$$h = \begin{pmatrix} x_1 \; x_2 \; x_3 \cdots x_n \\ y_1 \; y_2 \; y_3 \cdots y_n \end{pmatrix} \text{ and } m = \begin{pmatrix} y_1 \; y_2 \; y_3 \cdots y_n \\ z_1 \; z_2 \; z_3 \cdots z_n \end{pmatrix} \text{ then}$$

$$h * m = \begin{pmatrix} x_1 \; x_2 \; x_3 \cdots x_n \\ z_1 \; z_2 \; z_3 \cdots z_n \end{pmatrix} \text{ and } h^{-1} = \begin{pmatrix} y_1 \; y_2 \; y_3 \cdots y_n \\ x_1 \; x_2 \; x_3 \cdots x_n \end{pmatrix}$$

Thus, $(H,^*)$ is a group, called as the permutation group [1, 6].

Example 12.2 Show that all permutations on $S = \{3, \ 6, \ 8\}$ is a group with respect to multiplication of permutations.

Solution Let $H = \{I = d_1, \ d_2, d_3, d_4, d_5, d_6\}$, where,

$$d_1 = \begin{pmatrix} 3 & 6 & 8 \\ 3 & 6 & 8 \end{pmatrix}, \quad d_2 = \begin{pmatrix} 3 & 6 & 8 \\ 3 & 8 & 6 \end{pmatrix}, d_3 = \begin{pmatrix} 3 & 6 & 8 \\ 6 & 3 & 8 \end{pmatrix},$$

$$d_4 = \begin{pmatrix} 3 & 6 & 8 \\ 6 & 8 & 3 \end{pmatrix}, d_5 = \begin{pmatrix} 3 & 6 & 8 \\ 8 & 3 & 6 \end{pmatrix}, d_6 = \begin{pmatrix} 3 & 6 & 8 \\ 8 & 6 & 3 \end{pmatrix},$$

Consider the following table having elements of H and binary operation *.

*	d_1	d_2	d_3	d_4	d_5	d_6
d_1	d_1	d_2	d_3	d_4	d_5	d_6
d_2	d_2	d_1	d_4	d_3	d_6	d_5
d_3	d_3	d_5	d_1	d_6	d_2	d_4
d_4	d_4	d_6	d_2	d_5	d_1	d_3
d_5	d_5	d_3	d_6	d_1	d_4	d_2
d_6	d_6	d_4	d_5	d_2	d_3	d_1

i. Closure Property: Each element of the table belongs to H. Therefore, * is BO.
ii. * is associative.
iii. Identity Element $=d_1 = I$
iv. From the table, the following are the elements and the corresponding inverse.

Element	d_1	d_2	d_3	d_4	d_5	d_6
Inverse	d_1	d_2	d_3	d_5	d_4	d_6

Thus, $(H, *)$ is a permutation group.

v. This is not commutative as $h*m \neq m*h$. Therefore, $(H, *)$ is not Abelian group. It is the smallest non Abelian group [7].

12.6 Complexes and Subgroups: Experiential Learning

(a) **Brainstorming Experiment**

In Experiment-I, find the subsets of the set H. There are eight subsets of the set H, given as

$$H_1 = \phi,\ H_2 = \{M\},\ H_3 = \{F\},\ H_4 = \{S\},\ H_5 = \{M, F\},$$
$$H_6 = \{M, S\},\ H_7 = \{F, S\},\ H_8 = \{M, F, S\}$$

Now, test whether the above subset preserve all properties of the set or not. That is, test whether the above subsets are groups under the same operation or not.

(b) **Inferences**

The following are the significant inferences.

1. The subsets H_1, H_2, and H_8 groups, called as subgroups of the group $(H,*)$.
2. The remaining subsets are only subsets, but not groups. So, they are called as Complex of a group.

(C) Mathematical Insights

Complex of a Group

Any non-empty subset of a group (*H*, *) is a complex of a group.

For example, The sets $H_1 = \{1, 2, 3, 4, 5, 6\}$, $H_2 = \{2, 4, 6, \cdots\}$, and $H_2 = \{10\}$ are complexes of a group $(Z, +)$.

Subgroup (SG)

A set $S \neq \phi$ *and* $S \subseteq H$ is a subgroup of a group (*H*, *), if (*S*, *) itself is a group [8].

Examples

1. $(Z, +)$ is a SG of $(R, +)$.
2. $(2Z, +)$ is a SG of $(Z, +)$.
3. $(R,.)$ is not a SG of $(C,.)$.

Note:

Improper SG: $(\{e\}, *)$ and $(H, *)$
Proper SG: It is not Improper SG.
Theorems on Subgroups

Theorem 12.2 *The identity elements of the group (H) and subgroup (S) are same.*

Proof Let $e_1 \in H$ and $e_2 \in S$

Let $s \in S \Rightarrow\ s * e_1 = e_1 * s = s$.
Also, $s \in S \Rightarrow s \in H \Rightarrow\ s * e_2 = e_2 * s = s$
$\Rightarrow\ s * e_1 = s * e_2\ \Rightarrow\ e_1 = e_2$.

Theorem 12.3 *If* $(H, *)$ *and* $(S, *)$ *are groups and subgroups, respectively, such that* $S \subseteq H$ *and* $x^{-1} = y$ *in H and* $x^{-1} = z$ *in S then* $y = z$.

Theorem 12.4 $(S, *)$*is a subgroup of the group* $(H, *)$*iff* $h * d^{-1} \in S;\ \forall\, h, d \in S$.

Proof Suppose $(S, *)$ is a subgroup. Therefore,

$$h, d \in S\ \ \Rightarrow h, d^{-1} \in S \Rightarrow h * d^{-1} \in S.$$

Now, $h * d^{-1} \in S;\ \forall\, h, d \in S$. Claim: To prove that $(S, *)$ is a subgroup.

i. Closure Property: Let $h, d \in S\ \Rightarrow\ h, m^{-1} \in S\ \Rightarrow h * \left(d^{-1}\right)^{-1} = h * d \in S$.
ii. Associativity: $(S, *)$ is associative.
iii. Existence of the Identity Element: Let $h \in S\ \Rightarrow\ h, h \in S\ \Rightarrow h * (h^{-1}) = e \in S$.
iv. Existence of the Inverse Element: Let $e, h \in S\ \Rightarrow\ e*(h^{-1}) = h^{-1} \in S,\ \forall\, h \in H$.

Thus, $(S, *)$ is a subgroup of $(H, *)$.

Theorem 12.5 *Intersection of two Subgroups*

If $(S_1, *)$ *and* $(S_2, *)$ *are subgroups of the group* $(H, *)$ *then* $(S_1 \cap S_2, *)$ *is a subgroup.*

Theorem 12.6 *Union of two Subgroups*

If $(S_1, *)$ *and* $(S_2, *)$ *are subgroups of the group* $(H, *)$ *then* $(S_1 \cup S_2, *)$ *is a subgroup iff* $S_1 \subseteq S_2$ or $S_2 \subseteq S_1$.

Example 12.4 Is union of two subgroups is a subgroup? If not, give an example of it.

Solution The union of two subgroups is not necessarily a subgroup.

Let $S_1 = \{\cdots, -4, -2, 0, 2, 4, \cdots\}$ and $S_2 = \{\cdots, -6, -3, 0, 3, 6, \cdots\}$ are subgroups of the group $(Z, +)$. Now,

$$S_1 \cup S_2 = \{\cdots - 6, -4, -3, -2, 0, 2, 3, 4, 6 \cdots\}$$

$2, 3 \in S_1 \cup S_2$ but, $2 + 3 = 5 \notin S_1 \cup S_2$

Therefore, $S_1 \cup S_2$ is not closed with respect to $+$. Hence, $(S_1 \cup S_2, *)$ is not a subgroup.

12.7 Order of an element: Experiential Learning

Let $(H, *)$ be a group having identity element e. The smallest positive integer r is called the order of the element $h \in H$, if $h^r = e$ [9].

It is denoted by $o(h) = r$. If no such integer exists, then $o(h) = 0$ or $o(h) = \infty$.

The order of the group is the number of elements in that group.

Notes:

1. If $e \in H$ is the identity element in $(H, *)$, then $o(e) = 1$.
2. In group $(H, *)$, $o(d) = o\left(d^{-1}\right)$; $\forall d \in H$.
3. In group $(H, *)$, $o(d) \leq o(H)$; $\forall d \in H$.

12.8 Cyclic Group: Experiential Learning

A group $(H, *)$ is a cyclic group if $\exists\, g \in H$ s.t. $h = g^n$; $\forall\, h \in H,\ n \in Z^+$, The element g is a generator of $(H, *)$ and $H = \langle g \rangle$.

The generators of a group may be more than 1.

Example 12.5 Find the generators of the multiplicative cyclic group on $H = \{1, -1, i, -i\}$.

Solution $\therefore H = \langle i \rangle = \langle -i \rangle$.

Theorem 12.7 *Every cyclic group is an Abelian group.*

Proof Consider, $h * m = g^p * g^q = g^{p+q} = g^{q+p} = g^q * g^p = m * h$

Thus, $(H, *)$ is abelian group.

Theorem 12.8 *In cyclic group, If* $H = \langle g \rangle$ *then* $H = \langle g^{-1} \rangle$.

Proof We have, $h \in H \Rightarrow h = g^r$, r is some positive integer.

$\Rightarrow h = g^r = (g^{-1})^{-r}$. Thus, g^{-1} is also generator of H.

Theorem 12.9 *Lagrange's Theorem*

If S is a subgroup of the finite group $(H, *)$ *then* $o(S)$ *divides* $o(H)$.

Theorem 12.10 *If H is any element of the group* $(H, *)$ *then* $o(h)$ *divides* $o(H)$.

Proof Use division algorithm to prove this theorem.

Theorem 12.11 *If* $(H, *)$ *is a finite group of order n and* $d \in H$ *then* $d^n = e$.

Example 12.6 Construct a group that is Abelian but not cyclic.

Proof Let $H = \{I, A, B, C\}$ where,

$$I = \begin{bmatrix} 1 & 0 \\ 0 & 1 \end{bmatrix}, A = \begin{bmatrix} -1 & 0 \\ 0 & 1 \end{bmatrix}, B = \begin{bmatrix} 1 & 0 \\ 0 & -1 \end{bmatrix}, C = \begin{bmatrix} -1 & 0 \\ 0 & -1 \end{bmatrix}$$

H is Abelian w.r.t. matrix multiplication.

The identity element is I. So, $o(I) = 1$.

$$o(A) = o(B) = o(C) = 2 \quad \because A^2 = I = B^2 = C^2.$$

As there no element of order 4, so, H is not a cyclic group.

12.9 Normal subgroups

A subgroup N of a group $(H, *)$ is a normal subgroup of H, if for all $h \in H$ and for all

$$n \in N \quad h * n * h^{-1} \in H$$

Improper Normal SG: $(\{e\}, *)$ and $(H, *)$.

Simple Group: A group $(H, *)$ is a simple group if it has only two normal subgroups namely $(\{e\}, *)$ and $(H, *)$.

Example 1 $(Z, +)$ is an Abelian group. $(2Z, +)$, $(3Z, +)$, and $(13Z, +)$ are normal subgroups.

12.10 Quotient Groups

Let $(H, *)$ is a group and $(N, *)$ is a normal subgroup of $(H, *)$. Let H/N is the set all cosets of N in H.

$$H/N = \{N * h / h \in H\}$$

$(H/N, *)$ is called the quotient group or factor group.

12.11 Isomorphism of Groups: Experiential Learning

Homomorphism of Groups

Let $(F, *)$ and $(Z, \bullet)$ be two groups. A function $\lambda : (F, *) \rightarrow (Z, \bullet)$ is a homomorphism if

$\lambda (f * t) = \lambda(f) \bullet \lambda(t)$ for all $f, t \in F$, i.e., $f * t \in F$ & $\lambda(f) \bullet \lambda(t) \in Z$.

A homomorphism $\phi : (F, *) \rightarrow (F, *)$ is called an endomorphism [10].

The pictorial representation of the homomorphism of groups is depicted in the following figure.

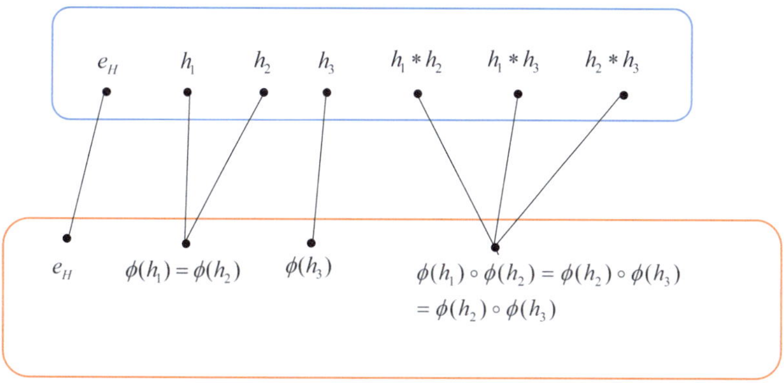

Theorem 12.12 *If a function $\phi : (H, *) \rightarrow (K, \bullet)$ is a homomorphism then*

i. $\phi (e_H) = e_K$ *where, e_H &e_K are the identity elements in H and K, respectively.*
ii. $\phi (h^{-1}) = [\phi (h)]^{-1}$

Proof

i. Consider, $\phi(h) \cdot e_K = \phi(h) = \phi(h * e_H) = \phi(h) \cdot \phi(e_H)$.

$$\phi(h) \cdot e_K = \phi(h) \cdot \phi(e_H) \Rightarrow e_K = \phi(e_H)$$

ii. We have, $e_K = \phi(e_H) = \phi\left(d * d^{-1}\right) = \phi(d) \cdot \phi\left(d^{-1}\right)$

$$\Rightarrow \phi\left(d^{-1}\right) = [\phi(d)]^{-1}$$

Isomorphism of Groups

Let $(H, *)$ and $(K, \cdot)$ be two groups. A function $\phi : (H, *) \rightarrow (K, \cdot)$ is said to be an isomorphism if

1. $\phi : (H, *) \rightarrow (K, \cdot)$ is homomorphism, i.e., $\phi(h * m) = \phi(h) \cdot \phi(m)$.
2. $\phi : (H, *) \rightarrow (K, \cdot)$ is bijective mapping, i.e., ϕ is one-one and onto.

Notes:

i. If $\phi : (H, *) \rightarrow (K, \cdot)$ is an isomorphism of groups then $(H, *)\&(K, \cdot)$ are called isomorphic groups and denoted by $H \cong K$.
ii. An isomorphism $\phi : (F, *) \rightarrow (F, *)$ is called as automorphism of group H.

Example 12.7 Show that $(H, +) \cong (K, \cdot)$, where

$H = \{\cdots, -2, -1, 0, 1, 2, \cdots\}$ and $K = \left\{\cdots, p^{-2}, p^{-1}, 0, p^1, p^2, \cdots\right\}$, where p is any non-zero integer.

Solution Let $(H, +)$ and $(K, \cdot)$ be additive and multiplicative groups. Let $\phi : (H, *) \rightarrow (K, \cdot)$ is defined as $\phi(h) = p^h$; $\forall\, h \in H$.

We have, $\phi(h + m) = p^{h+m} = p^h \cdot p^m = \phi(h) \cdot \phi(m)$.
Therefore, ϕ is a homomorphism.
Now, $\phi(h) = \phi(m) \Rightarrow p^h = p^m \Rightarrow h = m$, So, ϕ is one-one function.
For any $p^h \in K$, $\exists\, h \in H$ such that $\phi(h) = p^h$. So, the function ϕ is onto function.
Thus, $\phi : (H, *) \rightarrow (K, \cdot)$ is an isomorphism.

12.12 Group Codes

Let H_n be the set of all binary words of length n. Let $\oplus$ be a binary operation on H_n such that for all $h, m \in H_n$, where $h = (h_1, h_2, h_3, \cdots h_n)$ and $m = (m_1, m_2, m_3, \ldots, m_n)$ and $h \oplus m = (h_1 \oplus m_1, h_2 \oplus m_2, h_3 \oplus m_3, \ldots h_n \oplus m_n)$ [1, 6]

The binary operation $\oplus$ denotes the addition modulo 2 on $\{0, 1\}$ and defined as follows.

$\oplus$	0	1
0	0	1
1	1	0

The algebraic structure $(H_n, \oplus)$ forms a group in which, the identity element is 0 = (0,0,0,…,0) and each element is its own inverse. Any code which is a group under the operation $\oplus$ is called a group code. The group code was first coined by Hamming and it plays an important role in binary encoding techniques.

The weight of any word h in H_n is denoted by $w(h)$ and defined as

$w(h) =$ Number of one's in h

e.g. w(1, 1, 1, 1, 1) = 5, w(0, 1, 0, 1, 0) = 2, w(1, 1, 0, 0, 1) = 3

The Hamming distance between h and m is denoted by $d(h, m)$ and defined as

$d(h, m) =$ the number coordinates in h and m at which h_a and m_a are different.

e.g., Let $h = (1,1,0,0,1,0,1)$ and $m = (0,1,0,1,0,1,0)$. In h and m, the coordinates at the positions first, fourth, fifth, sixth and seventh are different. Therefore,

$$d(h, m) = 5$$

Moreover, $h \oplus m = (1, 0, 0, 1, 1, 1, 1)$ and $w\ h \oplus m = 5$. Hence, $d(h, m) = w(h \oplus m)$.

12.13 Case Studies and Projects

The following are case studies/projects of this chapter.

1. Applications of group theory to design cryptographic algorithms.
2. Group codes and error correcting codes.
3. Group theory algorithms to solve rubric cube.
4. Review on isomorphic groups and their applications
5. Role of algebraic structure with a binary operation in Applied Sciences (Physics, Chemistry, Mathematics, Medical)

Exercise

1. Test whether the following are groups or not.
   i. Set of real numbers with respect to addition.
   ii. Set of rational numbers with respect to multiplication.
   iii. Set of integers with respect to subtraction.
   iv. Set of complex numbers with respect to multiplication.
   v. Set of all 2×2 matrices with real entries and with respect to usual addition.
2. Let H be a group and $h, m \in H$. consider the equations $mx = h$ and $ym = h$. Show that
   i. $mx = h$ has a unique solution for $x \in H$.
   ii. $ym = h$ has a unique solution for $y \in H$.
   iii. Find an example of a group such that $x \neq y$ (i.e., Non-abelian group)
3. Prove that a group H is an abelian iff $(mh)^2 = m^2h^2$, for all $h, m \in H$.

4. Construct an example of a group H such that $(mh)^2 \neq m^2h^2$, for all $h, m \in H$.
5. List at least five subgroups of the following groups

   i. $(R, +)$
   ii. (Z, x)

6. Show that the following are subgroups:

   i. $\{1, i\} \subset (C, \times)$, where C is the set of Complex numbers.
   ii. $\{3^x/x \in Q\} \subset (R, \times)$, where R is the set of Real numbers.
   iii. $\{z/z^n = 1\} \subset (C, \times)$

7. Show that if M is a subgroup of H and P is a subgroup of M, then P is a subgroup of H.
8. Prove that subgroup of a cyclic group is a cyclic.
9. Show that the following groups are isomorphic

   i. $Z \cong 2Z$
   ii. $R \cong R^+$

10. List all homeomorphisms from Z_{20} to Z_{25}.
11. Write any three applications of group codes.
12. If M and P are subgroups of group H and P is normal, then show that $P \cap M$ is a normal subgroup of M.
13. Find all normal subgroups of S_4.
14. Show that a subgroup is normal iff it is the kernel of a homomorphism.
15. Construct two examples of the cyclic subgroups of a group.

References

1. Smith G, Tabachnikova O (2000) Topics in group theory. Springer Publication. 978-1852332358
2. Baumslag B, Chandler B (2020) Theory and problems of group theory. Schaum's Outline Series McGraw-Hill Book Company. 978-9389691382
3. Davvaz B (2015) A first course in group theory. Springer, Heidelberg. 9789811663642
4. Erciyes K (2021) Discrete mathematics and graph theory. Springer, Cham. https://doi.org/10.1007/978-3-030-61115-6
5. Rosen K, Krithivasan K (2021) Discrete mathematics and its applications, Mc Graw Hill India. ISBN-978930727353
6. Jiri M, Jaroslav N (2011) Invitation to discrete mathematics. 2nd edn. Oxford University Press. ISBN-9780198570424
7. Peter J (2000) Sets, logic and categories. Springer Publication. ISBN-9781852330569
8. Tremblay JP, Manohar R (2023) Discrete mathematical structures with applications to computer science. McGraw Hill Education. ISBN-9780074631133
9. Berkovich Y, Janko Z (2011) De Gruyter expositions in mathematics 56. Walter de Gruyter GmbH
10. Gilbert (2003) Rings and fields. Pure and applied mathematics

Chapter 13
Algebraic Structures with Two Binary Operations

Mathematics is the supreme judge; from its decisions there is no appeal-Tobias Dantzig

Abstract This chapter deals with algebraic structures with two binary operations and bridges between them. The chapter defines Ring and its types, subring, integral domain, field, homomorphism of rings and polynomial rings by using the "BEIMI" philosophy. Moreover, it includes precise examples and related theories for the better understanding of the readers. It also includes the similarities and differences among the algebraic structures connected by two binary operations. The chapter concludes with case studies, projects and exercises designed to bridge the gap between theoretical algebraic structures and their multidisciplinary applications.

Keywords Ring · Integral domain · Field · Ring homomorphism

13.1 Introduction

We have studied algebraic structures with one binary operation, such as Semigroup, Monoid, Group, Abelian group, Subgroup, and their types with suitable illustrations. Now, what will happen if there is a set with 2 or more binary operations? The similar questions are resolved in this chapter. This chapter deals with ring and its types, subring, integral domain, field, homomorphism of rings, and polynomial rings [1, 2].

13.2 Ring and Commutative Ring: Experiential Learning

(A) **Brainstorming Experiment-I**

Let us continue the real story given in the previous chapter. Let H be the set of all members in the family, $H = \{M, F, S\}$ and the operation $*$ defined as $h * m = p$ iff h and m are together then both help in the work of p or both think about p.

© The Author(s), under exclusive license to Springer Nature Singapore Pte Ltd. 2026
H. R. Bhapkar and P. N. Mahalle, *Discrete Mathematics: An Experiential Learning Perspective*, Studies in Autonomic, Data-driven and Industrial Computing,
https://doi.org/10.1007/978-981-95-7210-6_13

The tabular representation of the above story is as follows.

$*$	M	F	S
M	M	F	S
F	F	S	M
S	S	M	F

Now, define the other binary operation on the set H as $h \circ m = p$ iff h and m are together then both feel that p is a good cook. When M is with F or S, both feel that M is a good cook. If F and S are together, then they agree that S is a good cook. But when F or S are alone then feels that F is a good cook. The tabular representation of this is as follows.

$\circ$	M	F	S
M	M	M	M
F	M	F	S
S	M	S	F

Now, find out the precise inferences.

(B) **Inferences-I**

The following are a very special inferences of the above experiment.

1. An algebraic structure $(H, *, \circ)$ is an abelian group w.r.t. "$*$."
2. An algebraic structure $(H, *, \circ)$ is semigroup w.r.t. "$\circ$."
3. "$\circ$" distributes over "$*$." Therefore, $(H, *, \circ)$ is a ring.
4. An algebraic structure $(H, *, \circ)$ is semigroup and commutative with respect to "$\circ$." Therefore, it is a commutative ring.
5. An algebraic structure $(H, *, \circ)$ is semigroup and it has the identity element with respect to "$\circ$." Therefore, it is a ring with unity.

(C) **Mathematical Insights-I**

The following are very significant Mathematical insights of the above inferences. [1, 3]

13.2.1 Ring

A non-empty algebraic structure $(R, *, \circ)$ is called a ring, if it satisfies the following axioms.

1. $(R, *)$ is an Abelian group.
2. $(R, \circ)$ is Semigroup.

3. $' \circ '$ distributes over '*' (Second operation distributes over the first operation)

$$D_1.\ h \circ (a * f) = (h \circ a) * (h \circ f)$$
$$D_2.\ (h * a) \circ f = (h \circ a) * (h \circ f)$$

13.3 Types of Rings

In a ring structure there are two binary operations and a set. Moreover, there are only five properties of an operation such as binary operation, associativity, existence of the identity, existence of inverse, and commutativity. In a ring, the set follows all these five properties with respect to the first operation and the first two properties with respect to the second operation. Moreover, there is a distributive law for connecting two operations together. However, there is a scope to add remaining three properties w.r.t. the second BO that will motivate us to define the new types of rings [4].

13.3.1 Commutative Ring

A ring $(R, *, \circ)$ is called a commutative ring (CR) if $h \circ m = m \circ h\ ;\ \forall\ h, m \in R$.

13.3.2 Ring with Unity

A ring $(R, *, \circ)$ is called a ring with unity if $\forall\ h \in R,\ \exists\ \ 1 \in R$ such that $h \circ 1 = 1 \circ h\ = h$.

The unity is also known as the unit element [5].

Examples:

1. (Z, +, .) is a CR with unity, here, BO are " + " and "."
2. (2Z, +, .) is a commutative ring without unity, here 2Z = set of even integers.

Properties of a Ring

1. If $(R, *, \circ)$ is a ring with zero element e_0 and unity element e_1 then following are true for all $h, m, z \in R$.

   i. Zero element is unique.
   ii. $h \circ (-z) = (-z) \circ m = -(z \circ m)$.
   iii. $(-h) \circ (-z) = h \circ z$.
   iv. Unit element is unique.

These are the basic properties, so the readers can prove it by referring the previous chapter.

Note: The cancelation laws with respect to $\circ$ may or may not be true in the ring $(R, *, \circ)$ [1].

2. If $(R, *, \circ)$ is a ring, then
   i. The identity element (Zero element) with respect to $*$ is unique.
   ii. The inverse with respect to $*$ is unique.
   iii. If R has a unity (Identity element with respect to $\circ$) then it is unique.
   iv. If R has unity and h is the unit element in R, then the inverse of h with respect to $\circ$ is unique.

Example 13.1 Let $Z_n = \{1, 2, 3, \ldots, (n-1)\}$. Let $\oplus$ and $\otimes$ denote the BO of " + " modulo n and "." modulo n, respectively, defined as follows.

$$\begin{aligned} h \oplus m &= h + m; \text{ if } h + m < n \\ &= h + m - n; \text{ if } h + m \geq n \end{aligned}$$

and $h \otimes m =$ the remainder of $h\,m$ divided by n. Show that $(Z_n, \oplus, \otimes)$ is a ring. [6, 7]

Solution: We have, $(Z_n, \oplus)$ is an Abelian group with 0 as the identity element and $n - p$ is the inverse of p. Now, prove that $(Z_n, \otimes)$ is a Semigroup.

By the definition, Z_n is closed with respect to $\otimes$. Now, prove that $\otimes$ is associative on Z_n.

$h \otimes m =$ the remainder of $h\,m$ divided by n.
i.e., $h\,m = xn + a$. So, $h \otimes m = a$.
$(h \otimes m) \otimes p = a \otimes p = b$ as $ap = yn + b$

$$\begin{aligned} &\Rightarrow (hm - xn)p = yn + b \\ &\Rightarrow hmp = xnp + yn + b = (xp + y)n + b \end{aligned}$$

Now,
$m \otimes p =$ the remainder of mp divided by n.
i.e., $mp = zn + c$. So, $m \otimes p = c$.
$h \otimes (m \otimes p) = h \otimes c = d$ as $hc = qn + d$

$$\begin{aligned} &\Rightarrow h(mp - zn) = qn + d \\ &\Rightarrow hmp = hzn + qn + d = (hz + q)n + d \end{aligned}$$

So, $b = d$ and.
Therefore, $(Z_n, \otimes)$ is a Semigroup.
Let us prove that $\otimes$ distributes over $\oplus$.
Consider, LHS

$$m \oplus p = m + p \quad \text{or} \quad m \oplus p = m + p - n$$
$$\Rightarrow h \otimes (m \oplus p) = h(m + p) \bmod (n) \quad \text{or} \quad h(m + p - n) \bmod (n)$$

$$= (hm + hp) \text{ mod } (n) \quad \text{or} \quad (hm + hp - hn) \text{ mod } (n)$$
$$= hm \text{ mod } (n) + hp \text{ mod } (n) \quad \text{or} \quad hm \text{ mod } (n) + hp \text{ mod } (n) - n$$

Consider, RHS

$$h \otimes m = hm \text{ mod } (n) = s$$
$$h \otimes p = hp \text{ mod } (n) = r$$
$$\Rightarrow (h \otimes m) \oplus (h \otimes p) = (s + r) \text{ mod } (n)$$
$$= (s + r) \quad \text{or} \quad s + r - n$$
$$= hm \text{ mod } (n) + hp \text{ mod } (n) \quad \text{or} \quad hm \text{ mod } (n) + hp \text{ mod } (n) - n$$

Hence, LHS = RHS. Thus, $(Z_n, \oplus, \otimes)$ is a ring.

13.4 Subrings: Experiential Learning

Let $S \neq \varphi$ & $S \subseteq R$, S is a subring of the ring $(R, *, \circ)$ if $(S, *, \circ)$ is a ring [8].

Theorem 13.1 *A set* $S \neq \varphi$ & $S \subseteq R$, ***S is a subring of the ring*** $(R, *, \circ)$***iff***

1. $f * z \in S \,\&\, f \circ z \in S;\ \forall f, z \in S.$
2. $-z \in S\ ;\ \forall z \in S.$

Example 3 $(2Z, +, .)$ is a subring of the ring $(Z, +, .)$.

13.5 Integral Domain: Experiential Learning

(A) **Brainstorming Experiment-II**

Let's consider Experiment-I

(B) **Inferences**

The following are notable inferences of the above experiment.

1. An algebraic structure (H, ∗, ∘) is a commutative ring.
2. An algebraic structure (H, ∗, ∘) is without zero divisors with respect to "∘."

Thus, (H, ∗, ∘) is an integral domain.

(C) **Mathematical Insights**

Zero Divisors: Let $(R, *, \circ)$ be a commutative ring. A non-zero element $h \in R$ is said to be zero divisor if $\exists,\ m \neq 0 \in R$ such that $h \circ m = 0$ [9, 10].

A ring $(R, *, \circ)$ is said to be without zero divisors if $h \circ m = 0 \Rightarrow h = 0\ or\ m = 0;\ \forall\ h, m \in R$. Otherwise, the ring is known as the ring with zero divisors.

Examples

1. $\overline{2}$ is a zero divisor in $(Z_4, +, \cdot)$ as $\overline{2} \cdot \overline{2} = \overline{0}$.
2. $(M_{2\times 2}(\Re), +, \cdot)$ is a ring with zero divisors as

$$\begin{bmatrix} 1 & 0 \\ 0 & 0 \end{bmatrix}\begin{bmatrix} 0 & 0 \\ 0 & 1 \end{bmatrix} = \begin{bmatrix} 0 & 0 \\ 0 & 0 \end{bmatrix}, \text{ but both matrices are non-zero.}$$

Integral Domain

A commutative ring without zero divisors, is called as an integral domain. [11]

Examples

1. $(Z_4, +, \cdot)$ is not an integral domain.
2. $(M_{2\times 2}(\Re), +, \cdot)$ is not an integral domain as

$$\begin{bmatrix} 1 & 0 \\ 0 & 0 \end{bmatrix}\begin{bmatrix} 0 & 0 \\ 0 & 1 \end{bmatrix} = \begin{bmatrix} 0 & 0 \\ 0 & 0 \end{bmatrix}, \text{ but both matrices are non-zero.}$$

3. $(Z, +, .)$ is an integral domain.

13.6 Field: Experiential Learning

(A) **Brainstorming Experiment-III**

Let's consider Experiment-I

(B) **Inferences**

The following are notable inferences of the above experiment.

1. An algebraic structure $(H, *, \circ)$ is a commutative ring.
2. An algebraic structure $(H, *, \circ)$ is without zero divisors with respect to "$\circ$."

   Thus, $(H, *, \circ)$ is an integral domain.

3. Every non-zero element in $(H, *, \circ)$ has an inverse with respect to "$\circ$." Hence, it is a Field.

(C) **Mathematical Insights**

We aware that in the ring $(R, *, \circ)$, the unity or unit element means the identity element with respect to the second binary operation and zero element is the identity element with respect to the first binary operation.

Definition Let $(R, *, \circ)$ be a commutative ring with unity. $(R, *, \circ)$ is said to be a Field if every non-zero element in R has inverse with respect to $\circ$ in R. [1]

Examples $(\Re, +, \cdot)$, $(Q, +, \cdot)$, $(C, +, \cdot)$ are fields. But, $(Z, +, \cdot)$ is an Integral Domain but not a field as multiplicative inverse of an element does not exist in Z.

Theorem 13.2 ***If*** $(F, *, \circ)$***is a field then it is an integral domain.***

Proof Let $h, b \in F$ & $h \circ b = 0$. If $h = 0$ then the result is obvious. If $h \neq 0$ then h^{-1} exists in F. Then $h \circ b = 0 \Rightarrow h^{-1} \circ (h \circ b) = h^{-1} \circ 0 \Rightarrow b = 0$. Therefore, the field $(F, *, \circ)$ is without zero divisors. Thus it is an integral domain.

Theorem 13.3 $(Z_n, \oplus, \otimes)$***is a field iff n is a prime.***

Proof Let n be a prime and $0 < x < n$. Then, $\gcd(x, n) = 1$, so there exists two integers p and q such that $xp + nq = 1 \Rightarrow xp = (-np) + 1$. Therefore, $xp \equiv 1 \bmod (n)$ *OR* $[x][p] = 1$. So, $[x]$ *or* $\bar{x}$ is unit in $(Z_n, \oplus, \otimes)$. Thus, $(Z_n, \oplus, \otimes)$ is a field.

Conversely, if n is not a prime, then $n = x\,y$, where, $0 < x, y < n$. So, $[x] \neq [0]$ and $[y] \neq [0]$, but $[x][y] = [xy] = [n] = [0]$. Hence, it is not an integral domain. So, it cannot be a field.

Theorem 13.4 ***In*** Z_n***,*** $[h]$***is a unit iff*** $\gcd(h, n) = 1$.

Proof If $\gcd(h, n) = 1$, then by above theorem, it is a field and every element is unit. Conversely, let $[h] \in Z_n$ and $[h]^{-1} = [m]$. Therefore, $[hm] = [h]\,[m] = 1$. So, $hm \equiv 1 \bmod (n) \Rightarrow hm = 1 + nx$, for some $x \in Z$. So,

$$1 = hm + n(-x) \Rightarrow \gcd(h, n) = 1.$$

Note

1. We have, $h \,(\bmod p)$ is the remainder when h is divided by p. So, $x = h\,(\bmod p)$ if and only if $x(\bmod p) = h\,(\bmod p)$ in Z.
2. For any $n > 1$ & $n \in Z^+$, there are $\varphi(n)$ units and $n - 1 - \varphi(n)$ proper divisors of zero in Z_n.

Example 13.2 List all unit elements and proper divisors of zero in the following rings (1) $(Z_6, \oplus, \otimes)$, (2) $(Z_{11}, \oplus, \otimes)$.

Solution

(1) In Z_6, $[5] \otimes [5] = [1]$. Therefore, $[1]$ and $[5]$ are the only unit elements as integers 1 and 5 are relatively prime to 6. There are only $n - 1 - \varphi(n) = 6 - 1 - 2 = 3$ proper divisors of zero. So, the proper divisors are $[2]$, $[3]$, $[4]$.

(2) $(Z_{11}, \oplus, \otimes)$ is a field, so every non-zero element is unit. Thus, $[1], [2], [3], \cdots, [10]$ are units in the given ring. There is no any proper divisors of this ring [5].

13.7 Ring Homomorphism

Definition: Let $(R, *, \circ)$ and $(H, \oplus, \oplus)$ be two rings. A function $\varphi : R \rightarrow H$ is called a ring Homomorphism if for all $x,\ y \in R$,

1. $\varphi(x * y) = \varphi(x) \oplus \varphi(y)$
2. $\varphi(x \circ y) = \varphi(x) \otimes \varphi(y)$

Note:

1. The ring homomorphic function may or may not be onto.
2. The set H is the homomorphic image of R if the function φ is onto.
3. The ring homomorphic function may or may not be one to one.
4. $|R| \geq |H|$

Example 13.3 Let function $\varphi : (Z, +, \cdot) \rightarrow (Z_6, \oplus, \otimes)$ defined as $\varphi(h) = [h] = \bar{h}$. Then test whether φ is homomorphism or not.

Now, if $\psi : (Z, +, \cdot) \rightarrow (Z, +, \cdot)$ defined as $\psi(h) = 2h$. The function ψ is ring homomorphism as

$$\psi(x + y) = 2(x + y) = 2x + 2y = \psi(x) + \psi(y) \text{ and}$$
$$\psi(x \cdot y) = 2(x \cdot y) = 2x \cdot 2y = \psi(x) \cdot \psi(y).$$

The function ψ is not onto as odd integers do not have pre-image in the domain set. But, The function ψ is one to one.

The following are the significant observations by the above examples.

Ring Isomorphism: A ring homomorphism $\varphi : (R, *, \circ) \rightarrow (H, \oplus, \otimes)$ is called a ring isomorphism if φ is one to one and onto.

Note: The homomorphism deals with the similar structures whereas the isomorphism refers to the identical or similar structures [7, 8].

Theorem 12.6 ***If*** $\Gamma : (R, *, \circ) \rightarrow (H, \oplus, \otimes)$***is a ring homomorphism, then***

(a) $\Gamma(0_R) = 0_H$.
(b) $\Gamma(-x) = -\Gamma(x)$.
(c) $\Gamma(n \circ x) = n \otimes \Gamma(x)$.
(d) $\Gamma(x^n) = \{\Gamma(x)\}^n$.

Example 13.4 Let $R = \left\{\begin{bmatrix} h & m \\ m & h \end{bmatrix};\ h, m \in Z\right\}$ is a ring with respect to usual " + " and "." of matrices. The function φ defined as $\varphi\left\{\begin{bmatrix} h & m \\ m & h \end{bmatrix}\right\} = h - m$. Show that φ is a homomorphism.

Solution We have,

$$\varphi\left\{\begin{bmatrix} h & m \\ m & h \end{bmatrix} + \begin{bmatrix} p & q \\ q & p \end{bmatrix}\right\} = \varphi\left\{\begin{bmatrix} h+p & m+q \\ m+q & h+p \end{bmatrix}\right\}$$
$$= (h+p) - (m+q) = (h-m) + (p-q)$$
$$= \varphi\left\{\begin{bmatrix} h & m \\ m & h \end{bmatrix}\right\} + \varphi\left\{\begin{bmatrix} p & q \\ q & p \end{bmatrix}\right\}$$

And,

$$\varphi\left\{\begin{bmatrix} h & m \\ m & h \end{bmatrix} \times \begin{bmatrix} p & q \\ q & p \end{bmatrix}\right\} = \varphi\left\{\begin{bmatrix} hp+mq & hq+mp \\ mp+hq & mq+hp \end{bmatrix}\right\}$$
$$= (hp+mq) - (hq+mp) = (h-m) \times (p-q)$$
$$= \varphi\left\{\begin{bmatrix} h & m \\ m & h \end{bmatrix}\right\} \times \varphi\left\{\begin{bmatrix} p & q \\ q & p \end{bmatrix}\right\}$$

Thus, φ is homomorphism.

Example 13.5 Show that $R = \left\{h + m\sqrt{3} \;/\; h, m \in Z\right\}$ with respect to usual addition and multiplication is an integral domain but not a field.

Solution The solution is easy, so readers can prove it.

13.8 Polynomial Ring

Let R be an arbitrary ring and x is a variable (an indeterminate). Let $R[x] = \left\{f(x) = \sum_{i=0}^{\infty} r_i x^i / \; r_i \in R\right\}$. The $R[x]$ forms a ring w.r.t. " + " and "." of polynomials, called as the polynomial ring [11].

Note: Let $f(x) = \sum_{i=0}^{\infty} r_i x^i$ and $g(x) = \sum_{i=0}^{\infty} p_i x^i$, then

i. $f(x) = g(x) \Leftrightarrow r_i = p_i \;;\; \forall\, i \in Z^+ \cup \{0\}$.
ii. $f(x) + g(x) = \sum_{i=0}^{\infty} m_i x^i$, where $m_i = r_i + p_i \;;\; \forall\, i$
iii. $f(x) \times g(x) = \sum_{i=0}^{\infty} m_i x^i$, where $m_i = \sum_{i+j=n} r_i\, p_j$

Example 13.6 In a ring $(Z_8, \oplus, \otimes)$, find $a(x) + m(x) + s(x)$, $a(x) \times s(x)$, $m^2(x)$, if $a(x) = 4 + 3x + 2x^3$, $m(x) = 5x + 4x^2$, $p(x) = 2 + 5x$.

Solution We have,

$$a(x) + m(x) + s(x) = 4 + 3x + 2x^3 + 5x + 4x^2 + 2 + 5x$$
$$= 6x + 13x + 4x^2 + 2x^3 = 6x + 5x + 4x^2 + 2x^3$$

$$a(x) \times s(x) = (4 + 3x + 2x^3) \times (2 + 5x) = 8 + 26x + 15x^2$$
$$+ 2x^3 + 10x^4 = 2x + 7x^2 + 2x^3 + 2x^4$$
$$m(x) \times m(x) = (5x + 4x^2) \times (5x + 4x^2) = 25x^2 + 40x^3 + 16x^4 = x^2.$$

13.9 Case Studies and Projects

The following are the case studies/projects of this chapter.

1. Rings and its applications
2. Applications of finite fields in cryptography.
3. List all subrings of Z_n.
4. Differentiate among ring, integral domain, and field.
5. Design very special examples such that field but not integral domain.
6. Applications of polynomial rings in coding and decoding the secret codes.

Exercise

1. Determine whether the following structure are ring, commutative ring, ring with unity or not.
   i. The set of all 2 × 2 matrices with respect to usual addition and multiplication.
   ii. The set of all 3 × 3 diagonal matrices with respect to usual addition and multiplication.
   iii. $H = \left\{h + m\sqrt{5}\ /h, m \in Z\right\}$ with respect to ordinary addition and multiplication.
   iv. $(Z_{10}, \oplus_{10}, \otimes_{10})$
   v. $(Z_7, \oplus_7, \otimes_7)$
2. Show that the set of all 2 × 2 matrices of the form $\begin{bmatrix} 0 & x \\ 0 & y \end{bmatrix}$ is a subring of the ring given in example (1).
3. Find the multiplicative inverse of $\overline{55}$ in Z_{196}.
4. Find the multiplicative inverse of 21 in Z_{25}.
5. Prove that field can't have any zero divisors.
6. Prove that Z_p is a field iff p is prime.
7. Find the solution of each equation in Z_7.
   i. $x^2 + \overline{2}x + \overline{3} = \overline{4}$
   ii. $x^2 + \overline{4}x + \overline{1} = \overline{3}$
8. Show that the set of matrices of the form $\begin{bmatrix} p & q \\ 0 & r \end{bmatrix}$ is a subring of 2 × 2 matrices with integral elements.
9. Prove that every finite integral domain is a field. Is converse true. Justify it.

10. Every filed is an integral domain.
11. Explain any two applications of ring and field in real life.
12. Explain the role of ring and field in cryptography.
13. How field play an important role in AI and ML.
14. Explain ant two applications of polynomial rings in computer and security.
15. Give an example of a field which is not an integral domain.

References

1. Wallace DAR (1998) Groups, rings and fields. Springer. ISBN: 978-3540761778
2. Cohn PM (2012) Introduction to ring theory. Springer. ISBN: 978-1447104759
3. Davvaz B (2015) A first course in group theory. Springer. ISBN: 9789811663642
4. Loehr NA (2024) Advanced linear algebra. CRC Press
5. Anderson M, Feil T (2019) A first course in abstract algebra—rings, groups, and fields. 3rd edn. Chapman and Hall/CRC
6. Erciyes K (2021) Discrete mathematics and graph theory. Springer, Cham. https://doi.org/10.1007/978-3-030-61115-6
7. Rosen K, Krithivasan K (2021) Discrete mathematics and its applications. Mc Graw Hill, India. ISBN: 978930727353
8. Jiri M, Jaroslav N (2011) Invitation to discrete mathematics, 2nd edn. Oxford University Press. ISBN: 9780198570424
9. Tremblay JP, Manohar R (2023) Discrete mathematical structures with applications to computer science. McGraw Hill Education. ISBN: 9780074631133
10. Cohn PM (2000) Introduction to ring theory. Springer
11. Wang X, Xu G, Wang M, Meng X (2015) Rings and fields. Mathematical Foundations of Public Key Cryptography

Chapter 14
Conclusion and Future Outlook

Abstract The chapter explains the gist of the book, Discrete Mathematics: An Experiential Learning Perspective, and future outlooks. It enlightens how to bridge the gap between abstract mathematical concepts and their real-world applications. Through experiential learning, case studies, visualizations, and hands-on projects, each chapter transformed theory into understanding, reflection, and practice. Moreover, the chapter elucidates a journey from abstraction to application, from observation to innovation. With experiential learning as its core, it empowers learners to think logically, solve creatively, and contribute meaningfully to the ever-growing digital world.

14.1 Summary

This book, *Discrete Mathematics: An Experiential Learning Perspective*, was conceptualized to bridge the gap between abstract mathematical concepts and their real-world applications. Through experiential learning, case studies, visualizations, and hands-on projects, each chapter transformed theory into understanding, reflection, and practice.

Beginning with the foundational question—*Why Discrete Mathematics?*—we explored its philosophical depth, necessity in modern science and engineering, and wide-reaching applications. The book then introduced the concept of experiential learning and pedagogies that emphasize learning by doing, collaboration, reflection, and innovation.

Chapters 3–13 systematically unfolded the essential topics of discrete mathematics:

- **Sets, Relations, and Functions** laid the groundwork for mathematical structures and logic.
- **Logic, Proofs, Counting Principles, and Probability** developed analytical reasoning and decision-making skills.
- **Graphs and Trees** demonstrated how networks, connectivity, and hierarchical structures power computer science, communication, and data organization.

© The Author(s), under exclusive license to Springer Nature Singapore Pte Ltd. 2026
H. R. Bhapkar and P. N. Mahalle, *Discrete Mathematics: An Experiential Learning Perspective*, Studies in Autonomic, Data-driven and Industrial Computing,
https://doi.org/10.1007/978-981-95-7210-6_14

- **Generating Functions, Recurrence Relations, and Algebraic Structures** offered powerful tools for algorithm analysis, cryptography, coding theory, and system modeling.

Each concept was reinforced through experiential learning modules—Venn diagrams, logic puzzles, graph coloring problems, combinatorics via real-life case studies, and algebraic structures seen in encryption, group theory, and modular arithmetic.

By integrating technology, engineering applications, real-case observations, and project-based learning, the book achieved its goal of making discrete mathematics not just a subject to study, but an experience to live.

14.2 Research Openings

Despite the depth of coverage, discrete mathematics continues to evolve with emerging technologies. The following are potential research directions inspired by this book:

(A) Graph Theory and Network Science
- Dynamic graph models for social networks, IoT systems, and cybersecurity.
- Graph algorithms in AI for knowledge graphs and recommendation systems.
- Quantum graph models and their computational complexities.

(B) **Algebraic Structures in Cryptography and Blockchain**
- Advanced group theory, ring structures, and lattice-based cryptosystems.
- Homomorphic encryption and post-quantum cryptography.
- Blockchain consensus algorithms using discrete mathematical frameworks.

(C) **Machine Learning and Discrete Mathematics Integration**
- Optimization of neural networks using combinatorics and graph theory.
- Boolean functions and decision trees in interpretable AI.
- Tensor networks, group theory, and symmetry-based learning.

(D) **Experiential Learning in Mathematics Education**
- Development of AR/VR-based tools for teaching discrete structures.
- Gamification-based pedagogy for graphs, logic circuits, and automata.
- Comparative studies on skill development using experiential vs. traditional methods.

(E) **Computational Modeling and Simulation**
- Modeling epidemics using graph theory and probability.
- Discrete event simulations in logistics, transportation, and supply chain.
- Automata and cellular systems for intelligent machines and robotics.

14.3 Future Outlook

Discrete mathematics will remain the backbone of digital innovation. The future will witness its deeper integration into:

- **Education**
- Curriculum transformation supported by experiential learning, project-based assessments, and interdisciplinary teaching intersecting AI, cybersecurity, and data science.
- Global adoption of virtual labs, remote collaboration platforms, and AI-based tutors for personalized learning of discrete concepts.
- **Technology and Industry**
- Discrete structures will drive advancements in AI, robotics, quantum computing, blockchain, cybersecurity, bioinformatics, and smart cities.
- Graph neural networks, algebraic topology in data analysis, and combinatorial optimization will shape intelligent systems.
- **Ethics and Human-Centric Innovation**
- As algorithms influence society, discrete mathematics will guide ethical frameworks for decision-making, privacy, and fairness in AI models.
- Responsible innovation will require mathematical transparency, verifiability, and reasoning.
- **Lifelong Learning Perspective**

Students, educators, and researchers are encouraged to see discrete mathematics not as a subject that ends with examinations, but as a language of reasoning, creativity, and innovation that evolves with time.

In conclusion, this book has been a journey from abstraction to application, from observation to innovation. With experiential learning as its core, it empowers learners to think logically, solve creatively, and contribute meaningfully to the ever-growing digital world.